T0258556

# WHAT IS LIFE?

# What Is Life?

## THE INTELLECTUAL PERTINENCE
## OF ERWIN SCHRÖDINGER

*Hans Ulrich Gumbrecht,*
*Robert Pogue Harrison,*
*Michael R. Hendrickson,*
*Robert B. Laughlin*

STANFORD UNIVERSITY PRESS

*Stanford, California*

Stanford University Press
Stanford, California

*What Is Life?* was originally published in German under the title
*Geist und Materie—Was ist Leben? Zur Aktualität von Erwin Schrödinger*
© Suhrkamp Verlag Frankfurt am Main 2008.

Library of Congress Cataloging-in-Publication Data

Geist und Materie. English
    What is life? : the intellectual pertinence of Erwin Schrödinger /
Hans Ulrich Gumbrecht ... [et al.].
        p.    cm.
    "Originally published in German under the title Geist und
Materie."
    Includes bibliographical references.
    ISBN 978-0-8047-6915-0 (cloth : alk. paper) —
    ISBN 978-0-8047-6917-7 (pbk. : alk. paper)
    1. Schrödinger, Erwin, 1887–1961—Philosophy. 2. Physics—
Philosophy. 3. Life (Biology)—Philosophy. 4. Philosophy and
science. 5. Physicists—Austria—Biography. I. Gumbrecht, Hans
Ulrich. I. Title.
QC16.S265W4313      2011
530.01—dc22                                    2010029914

Typeset at Stanford University Press in 11/14.5 Bembo

*Frontispiece:* Erwin Schrödinger in 1923.
Photo: Carl Wolf. The Granger Collection, New York.

# Contents

*The Authors*        vii

Introduction: The Sustainability of
Erwin Schrödinger's Thought        I
HANS ULRICH GUMBRECHT

Schrödinger on Mind and Matter        13
ROBERT POGUE HARRISON

Schrödinger's Trouble: How Quantum Mechanics
Got Created with a Logical Loose End        33
ROBERT B. LAUGHLIN

Exorcizing Schrödinger's Ghost: Reflections on
'*What Is Life?*' and Its Surprising Relevance to
Cancer Biology        45
MICHAEL R. HENDRICKSON

Keeping the Singular, Risking Openness:
Erwin Schrödinger's Way of World Experience        105
HANS ULRICH GUMBRECHT

*Notes and References*        123

# The Authors

HANS ULRICH GUMBRECHT is Albert Guérard Professor in Literature at Stanford University. Among his books on literary theory and literary and cultural history are *Eine Geschichte der spanischen Literatur* (1990); *Making Sense in Life and Literature* (1992); *In 1926—Living at the Edge of Time* (1998); *Vom Leben und Sterben der großen Romanisten* (2002); *The Powers of Philology* (2003); *Production of Presence* (Stanford, 2004); *In Praise of Athletic Beauty* (2006); and *California Graffiti—Bilder von westlichen Ende der Welt* (2010). Forthcoming are titles on '*Stimmung*' (mood, climate) and the post-1945 era as a time of 'latency'. Gumbrecht is a regular contributor to the Humanities section of the *Frankfurter Allgemeine Zeitung*, *NZZ* (Zürich), and *Estado de São Paulo*. He is a member of the American Academy of Arts and Sciences, Professeur attaché au Collège de France, and has been a visiting professor at numerous universities worldwide. In 2009–2010 he was a fellow of the Siemens Foundation in Munich.

ROBERT POGUE HARRISON is Rosina Pierotti Professor in Italian Literature and chair of the Department of French and Italian at Stanford University, where he has taught since 1985. His most recent book, *Gardens: An Essay on the Human Condition* (2008), explores the connection between culture and cultivation by tracing an aesthetic, literary, and philosophical history of gardens. He is also the author of *The Body of Beatrice* (1988); *Forests: The Shadow of Civilization* (1992); *Rome, la Pluie: A Quoi Bon Littérature?* (1994); and *The Dominion of the Dead* (2003). His books have been published in French, Italian, German, Japanese, and Korean. In 2005, Harrison started a literary talk show on KZSU radio called *Entitled Opinions*, which features conversations with a variety of scholars, writers, and scientists.

MICHAEL R. HENDRICKSON is the director of Surgical Pathology at Stanford University Medical Center. He graduated from the University of California at Los Angeles with a combined major in chemistry and mathematics, received his M.D. at Stanford University Medical School, and continued his postgraduate training at that institution. He joined the Pathology Faculty in 1975 and became full professor in 1994. He spent a sabbatical year at Oxford University studying biostatistics. His chief area of expertise is in gynecological cancers, their diagnosis, and management. He is coauthor of *Surgical Pathology of the Uterine Corpus* and numerous publications in gynecologic pathology. In addition, he has had a career-long interest in classificatory issues in oncopathology and is currently completing a book concerned with the philosophy of classification, diagnosis, and decision-making in surgical pathology. This work draws on aspects of molecular oncology, multivariate statistics, biological systematics, judgmental psychology, and the philosophy of biology.

ROBERT B. LAUGHLIN was born in Visalia, California, in 1950. He is the Anne T. and Robert M. Bass Professor of Physics at Stanford University. After receiving a B.A. in mathematics from the University of California at Berkeley, he completed a Ph.D. at MIT in 1979 and has taught at Stanford University since 1985. He has received numerous awards, including the E. O. Lawrence Award for Physics (1985) and the Oliver E. Buckley Prize (1986), and was a corecipient of the Nobel Prize for Physics in 1998. He is a fellow of the American Academy of Arts and Sciences, the National Academy of Sciences, and the American Association for the Advancement of Science. His current research is primarily in high-temperature superconductivity theory. His publications include *A Different Universe: Reinventing Physics from the Bottom Down* (2005); and *The Crime of Reason: And the Closing of the Scientific Mind* (2008).

# WHAT IS LIFE?

# Introduction

## The Sustainability of Erwin Schrödinger's Thought

HANS ULRICH GUMBRECHT

Erwin Schrödinger's eminent place in the history of the natural sciences is undisputed. As Robert Laughlin remarks in this volume, the theory of wave mechanics for which Schrödinger was awarded the Nobel Prize for Physics in 1933 constitutes the one "mathematical description of matter [that] we use today for everything from chemistry to subnuclear particles," not least because it transforms quantum mechanics "from a nightmare of confusing mathematics to something genuinely elegant." But from the current perspective, even the description of wave mechanics is not Schrödinger's most meaningful achievement. Michael Hendrickson demonstrates in his detailed analysis that a collection of speculations, concepts, and metaphors from a 1943 series of lectures by Schrödinger (published one year later under the title *What Is Life?*) was in fact a decisive, if not *the* decisive, impetus for the emergence of biogenetics (for example, the concept of the 'genetic code'), although Schrödinger himself never advanced very far in his knowledge of chemistry. Both facts—quantum mechanics and biogenetics—have long become common knowledge, though the theses, arguments, and equations to which they refer remain comprehensible only to those trained in the sciences.

However, facts like the ones this volume seeks to clarify and confirm are surely not all that constitutes Erwin Schrödinger's timeliness, which concerns us here. For facts alone do not account for the extent to which reading certain of his texts, above all 1944's *What Is Life?* and 1956's *Mind and Matter*, could generate such intellectual potential for our time, far exceeding Schrödinger's historical impact and lasting significance in at least two fields of natural science. Still, the basis of Schrödinger's intellectual timeliness does lie primarily in his objective achievements—those bits of knowledge that are the fruits of his labor—especially when we take into account how astonishingly resilient these achievements have proven in comparison to the merciless slide into obsolescence that is the usual fate of scientific knowledge. For example, it has become possible, thanks to new technologies of measurement, to conclude that in the historic debate among quantum physicists, Schrödinger has proven correct against Niels Bohr and Werner Heisenberg, although to the present day it is evidently considered a faux pas to point this out among scientists. Meanwhile, in the field of biogenetics, though Schrödinger's "account of the centralized agency of the all-powerful gene" has certainly been significantly revised, at the same time, almost paradoxically, other concepts, metaphors, and views, which he—speculatively—introduced into the debates of his time (such as the idea of a 'network' or that of the 'self-organizing system'), have catalyzed a dynamic development that has left behind the early phases of the pertinent research.

Indeed, these exceptional instances of enduring achievements within the natural sciences can at most awaken a remote admiration among those interlopers, the philosophically minded intellectuals; such staying power cannot explain the complex fascination and *Auseinandersetzung* at work when we encounter Schrödinger's texts. Singular historical achievements do not, for example, answer why three months of reading Schrödinger's essays followed by an associated symposium remains one of the most productive experiences of Stanford University's Philosophical Reading Group, a workshop open to students and faculty from all disciplines, which has been meeting for nineteen years. This volume hearkens back to the discussions of Schrödinger by the Philosophical Reading Group, despite the fact that the texts it contains do not stem from those earlier debates; nor in fact are they even thematically predetermined by them. How is it that a circle composed primarily of humanists, approaching as dilettantes the text of a scientist, indeed one of the great natural scientists, was able to reap intellectual profits as significant as those it would gain in reading Hannah Arendt's *The Human Condition*, or even Kant's *Critique of the Power of Judgment*—texts which bookended the Schrödinger readings in the program of the reading group?

The obvious answer that persists in the minds of the Philosophical Reading Group's participants is that Erwin Schrödinger's life and work (and here I use the somewhat well-worn formula "life and work" advisedly) invited us, by means of an outstanding example of the scientific capacity for innovative approaches to general questions, to ask after the conditions which make such innovations possible. Naturally, this is a problem whose historical significance is outbid by its relevance for the scientific community and policy of the present and future. The surprising, almost bizarre dominance of biographical answers that one initially comes across (question 1, below) points to the ultimately rather philosophical—in any case, no longer primarily scientific-historical or biographical—problem of the sustainability of Schrödinger's thought. However, from a philosophical

*Hans Ulrich Gumbrecht*

perspective we want nothing less than to ask, Did Schrödinger's way of thinking make it possible to examine the many facets of a problem that had up to that point remained obscure, and so give expression to that which, once identified, would redefine the outlook of each subsequent generation of researchers and thinkers?

This second problem, which is the focus of our introduction, considers the intrinsic conditions for the sustainability of Schrödinger's thought, the core of the persistence of his work; it inquires into the source of his essential timeliness—and no longer merely the external circumstances of it. This also suggests why Schrödinger is profitable reading for more than just scientists. The search for this core once more splits the problem into two: the first part concerns the style of Schrödinger's thought (question 2); insofar as we manage to grasp this way of thinking, it becomes possible to uncover the essence of its content (question 3).

*Question 1.* The apparent institutional and biographical conditions for the specificity of Schrödinger's thought are, in their convergence, enough to generate a certain amount of skepticism concerning the efficiency of large-scale (and necessarily cost-intensive) underwriting of research. Similar issues arise when one examines the external conditions for the achievements of Albert Einstein, who was Schrödinger's friend as well as his colleague.

The same text that was so instrumental in the emergence of biogenetics makes visible Schrödinger's most earnest—if not his most expert—philosophical endeavors. There were long phases in Schrödinger's life, especially the years after the end of the Great War, in which he spent most of his working hours reading philosophical texts, above all the works of Schopenhauer and collections of Indian proverbs. But he was also interested in contemporary philosophical positions, for instance, the work of his teacher Ernst Mach and the phenomenology of Edmund Husserl. Schrödinger drew on these readings with a certain amount of freedom, since he was under no pressure to consider himself a specialist. Both as a theoretician and in his rare experimental research he was solitary. It is certainly true that in his lectures, he captivated his listeners through stirring rhetorical brilliance, conceptual clarity, and descriptive precision, but it was nonetheless clear to him that he could hardly be productive as a member of a research group. Neither could he see himself as a successful tutor for advanced students. Erwin Schrödinger enjoyed civil and collegial relationships with all of the other capacities of that great age of physics, without restricting himself out of a sense of group solidarity to this or that cadre of collaborators or allowing himself to be constrained to circulate his discoveries only within the confines of a particular group.

The most natural scientific lifestyle for his type of work, and for himself, was the role of Fellow, the long-term, far-reaching position of independent guest at a university or research institution, a position that was offered to him multiple times between 1933 and 1956 in England and Ireland. Schrödinger was not interested in wealth but in securing his economic independence. He kept his distance from political ideologies or the political parties of his time just as much as from academic groups and schools, although he tended toward decisive opinions in science throughout his life. Without being demanding, Schrödinger expected that his needs would be met and his fancies realized—as might be expected (to take up for a moment the psychological perspective) from someone who grew up (like Theodor W. Adorno) as an only child doted on by an unmarried aunt. All these external conditions of his life and work coalesced to give Schrödinger an astonishing degree of independence from institutions and the social environment, whose contact and stimulation he nevertheless needed. Such a convergence of professional prudence and individual talent in a single researcher could neither then nor today be acquired but rather must always already be in existence—and then, in individual cases, be individually employed.

*Question 2.* It is hardly possible to determine whether it was a consequence of biographical circumstances, but in any case Erwin Schrödinger remained the kind of scientist who is active in many fields and areas of concern rather than developing his skills along a permanent line of research:

> In my scientific work (just as elsewhere in my life) I have never tarried very long in pursuing a single program or axis of thought that defined me. Although I do not collaborate well with colleagues, nor unfortunately with students, my work has not progressed independent of them, for interest in a question comes to me only after others already have a like interest. Only rarely do I have the first word; much more often I take my inspiration from a desire to contradict or correct someone and am most concerned with the important discoveries that follow on an initial thought.

When, in 1921, he was appointed to the University of Zürich, "the multifaceted nature of Schrödinger's work in the areas of mechanics, optics, magnetism, radioactivity, the theory of gravity, and acoustics [stood] in the foreground" for his future Swiss colleagues. Additionally, he was able "to hold the biometrics lectures the biology students clamored for." And after his rather late breakthrough—in comparison to the other greats of natural science—in 1926 with the work on wave mechanics, the motif of 'multifacetedness' dominated his profile when he was named as the second choice for a successor to Max Planck in Berlin, a position Schrödinger would accept a year later:

He has been well-known for several years for the multifaceted
nature of his work, and likewise for his powerful and profound
style in the identification and development of new issues in
physics from unexpected perspectives, and he has mastered
the whole spectrum of physical and mathematical methods.
Schrödinger is an excellent orator and interlocutor, which he
demonstrates through a straightforwardness and clarity, which his
charming South-German temperament only underscores.[1]

The combination of scientific 'multifacetedness' and 'depth'
emphasized here is at once more abstract and perhaps more
exact than the pronounced convergence, which one observes
again and again in Schrödinger's texts, between the identification of singular phenomena and the, in principle, inconclusive
and open-ended series of *interpretations* of these phenomena,
which invariably refer back to the respective phenomena again.
Such a convergence was shaped through the divergent influences of his early teachers, the experimental researcher Ludwig
Boltzmann and the epistemologist of science Ernst Mach. This,
in any case, allows us to conceive as an intriguing intellectual
gesture how Schrödinger's work, as Robert Harrison demonstrates, was driven by two fundamentally different passions: the
desire to explain something and the desire to reveal something.
The passion to explain leads to scientific knowledge, which thus
usually marks the culmination and endpoint of every investigation, whereas Schrödinger's research began with knowledge and
ended in awe.

This capacity, indeed this passion for inexhaustible wonder,
distinguished Schrödinger's intellectual style. His wonder grew
manifestly each time he reverted from the interpretation of a
phenomenon or from a corresponding speculation to the phenomenon itself. It was then that he was able to bracket his pres-

ence as observer, to step back to give the phenomenon the space needed to show itself, to be able to reveal its nature. Conventional scientific knowledge, by contrast, absorbs the phenomenon in mathematical formulas and freezes it.

Schrödinger's wonder in the presence of the phenomenon led him to return over and again to the development of new interpretations and hypotheses; he did not, to all appearances, attempt to predict the outcome of this process—he merely sought it. And because of this, he not only developed a particular receptiveness to those questions that led to no answer, no 'scientific knowledge'; he also increasingly concentrated on examining such questions as 'mysteries' instead of avoiding them or, worse, distorting them to fit some familiar pattern of explanation. Harrison calls this attitude intellectual honesty. It is certain that Schrödinger's honesty and the intellectual dynamic stemming from it constitute an essential source of the fascination generated in reading his texts and of that which is so often and so justly praised as his "intellectual elegance."

*Question 3.* This intellectual style alone, with its characteristic complexity and energy, was enough to distinguish Schrödinger as one of the great scientific authors. But this style, independent of any particular constellation of issues, does not explain the persistent relevance of certain of his texts. For there are certainly formidable, elegant, and flexible thinkers who have never earned Schrödinger's prestige and status. We must ask then whether Erwin Schrödinger's style of thought has led once again to inconclusive or at least yet-to-be-concluded problems. Again, I align myself here with Harrison's view that to these layers of phenomena belong those very "antinomies" that Schrödinger tackled in *Mind and Matter.* From his reflection on the relation between scientific observation and phenomena he arrived at the question of the "place where mind encounters matter." This was one of those moments where Schrödinger was "honest" enough not to force himself to produce an unequivocal answer that he didn't have. Notwithstanding, the problem did not loosen its hold on him but provoked the risky, philosophical, and particularly interesting conjecture that such a space of encounter between mind and matter might not exist at all.

Yet another problem without a solution was already evident in Schrödinger's earlier texts and would likewise preoccupy him to the end of his life. It concerns the possibility—so critical for him—that the soaring heights of interpretation and speculation would return, time and again, to a given phenomenon as his point of departure. He questioned how it was possible that the seemingly infinite variety of representations and interpretations could so often be subsumed in the vanishing point of the phenomenon. Of course, this is the classic epistemological problem of perspectivism. In his youth, Schrödinger apparently believed that Schopenhauer's philosophy offered a solution, which was derived from Indian thought: the teaching that all individual consciousnesses are merely fragments of a single transcendental consciousness. Whether in later years Schrödinger no longer found this answer wholly satisfying is difficult to guess. But he faced the question many times, without forcing himself to close off the exploration by positing a definitive solution.

From the standpoint of epistemological history, these two central questions—or rather, since both remain hitherto unanswered, these two "mysteries"—lead one back to the moment in the early nineteenth century that marked the emergence of the second-order observer.[2] Schrödinger's own intellectual preoccupations drew him further into them, but the questions were already there before him, independently ("objectively," so to speak) given. From the moment that for the 'subject' of early Western modernity—the world-observing Cartesian 'cogito'—

the practice of observing oneself in the act of observation became inevitable, it has likewise become exceedingly, and per-
sistently, clear that we do not know how to reconcile world
appropriation through the senses (perception) and world appropriation through concepts (experience). Schrödinger's search for
the place where mind meets matter is a variant of this problem.
Second, it has become apparent that for each phenomenon there
can be as many representations as there are observer perspectives,
which begs the question of whether we, in light of a potential
infinity of representations, can any longer believe in the existence of individual reference phenomena. This is the problem
that the young Schrödinger—through recourse to Schopenhauer and Indian philosophy—believed himself to have solved.

Since the early nineteenth century, scientists and philosophers were persuaded, time and again, to find definite solutions
and thus to eliminate at least one of these problems. Philosophy
of history from the Hegelian perspective, and evolutionism from
the Darwinian perspective, were for the better part of a century
considered solutions with regard to perspectivism; a similar status attached to the theory of relativity with regard to the question of the compatibility between conceptual and sensible world
appropriation. Erwin Schrödinger made crucial discoveries and
sustained productive institutions, but not by simply bracketing
out these inconclusive questions. In other words, the things we
know, thanks to Schrödinger, make us ever more conscious of
what we yet do not know. For in spite of the wealth of solutions
that present themselves, the world remains irreducibly complex;
thus, in Schrödinger's works we see each new solution giving
rise to a more acute awareness of the unresolved. And so we
arrive at the reason that Schrödinger's speculations could never
come to an end. Should we, however, find an answer to every
question left unsolved by Schrödinger and by those who suc-

ceeded him, that would represent the end, indeed the nullification, of the distinctive sustainability of Schrödinger's thought. But this loss would be the condition for the most substantial and most sought-after scientific advancement imaginable.

*Translated by Lisa Ann Villareal*

# Schrödinger on Mind and Matter

ROBERT POGUE HARRISON

In addition to being one of the great scientists of his age, Erwin Schrödinger was also one of its great thinkers. When they are great, both scientists and thinkers show us the world as it has never been seen before, disclosing truths that had not been accessible before. I would claim, even more radically, that their view of the world transforms the very modes of the world's self-manifestations. Yet these close cousins—scientists and thinkers—are driven by two fundamentally different passions: the passion for explanation in one case, and the passion for revelation in the other. Schrödinger, exceptionally, was driven by both. As a scientist he sought to explore and expound the laws of nature; as a thinker, to uncover the mystery of the phenomenal world over which those laws hold sway—the mystery of its provenance, its lawfulness, and above all its self-display to human witness. Every great thinker is a mystic of sorts, whereas every great scientist is a detective of sorts. Schrödinger was a detective who followed clues to the edge of science's field of vision, from which edge he peered into the *mysterium* of spiritual reality, which is intimately bound up with, yet strangely unlike, the reality of matter.

I am tempted to say that Schrödinger reversed the usual trajectory of scientific inquiry. Science begins in wonder and ends

13

in insight, whereas many of Schrödinger's inquiries began with insight and ended in wonder. A successful scientific investigation is supposed to put an end to wonder. That is why Francis Bacon, a classic apologist for modern science, denigrated wonder as an inappropriate, or at best temporary, scientific disposition. Bacon spoke of wonder as "broken knowledge" and as "contemplation broken off, or losing itself," by which he meant that as science advances toward "certainty," it dispels the initial ignorance that nourishes wonder. I invoke Bacon here not because I believe he was a great thinker but because his ideas about wonder as broken knowledge express a far from obsolete presupposition of scientific inquiry, namely that the unknown is merely as-yet unconquered territory and that what Bacon confidently called the "advancement of learning" dissipates the cloud of mystery that otherwise attends phenomena.[1]

For Schrödinger this was exactly *not* the case. If science seeks to explain natural phenomena by natural causes, he found in scientific explanation an intensification, expansion, and justification of wonder. The known is most wondrous of all; reality itself makes the case for mysticism. If one is a thinking scientist like Schrödinger, the natural world *in its intelligibility* is what proves astonishing. Far from demystifying the phenomenon, science renders it uncanny, provided we are ready to think about, and *think through*, what science has succeeded in explaining.

One of the best examples of the way Schrödinger thought through scientific explanation, enlisting it on behalf of wonder, is his little book *What Is Life?*[2] Here he looks at "the physical aspect of the living cell" from the point of view of the laws of physics, only to conclude that the reproductive patterns of living things so defy those laws that the physicist is obliged to acknowledge the bewildering exceptionalism of life in the order of matter. Schrödinger reminds us that the laws of physics are statistical in nature. With inanimate matter (and most of the universe is inanimate) there is such a quotient of irregularity in the motion of individual atoms that order can occur only in large statistical numbers. That is why atoms are so small—as Schrödinger explains in a beautiful chapter of *What Is Life?* entitled precisely "Why Are Atoms So Small?"—for it takes an almost incalculable quantity of atoms to overcome the statistical probabilities that defy order in even the smallest organizations of inanimate matter. "In biology," however, "we are faced with an entirely different situation"—a situation in which "a single group of atoms [that is, chromosomes] existing only in one copy produces orderly events, marvelously tuned in with each other and with the environment according to most subtle laws" (79). These laws, Schrödinger adds, "cannot be reduced to the ordinary laws of physics," for the transgenerational stability and resiliency of a single molecule containing all the genetic information for the reproduction of life is, from the point of view of physical law, so improbable as to be essentially miraculous. This is not to say that life does not obey natural laws. Schrödinger insists that it does, even if they are so different from those of physics as to be completely baffling, all the more so as science further succeeds in *explaining* the hereditary mechanisms by which life reproduces its forms in unimaginably fine detail.

I will leave it to my scientific colleagues in this volume to probe in depth the complexities of Schrödinger's investigation of life, his definition of the chromosome as an "aperiodic crystal," and the conclusions he reaches on the role that "negative entropy" plays in the sustenance of life.[3] What I would insist on here is that the marvel he expresses before the phenomenon of life in the final chapters of his book is *not* the breaking off of knowledge, nor is it mere puzzlement before the unknown. It is precisely to the degree that Schrödinger *comprehends* the mechanisms of life that he is in awe of them. That awe is an act of thinking, one that does not merely observe the phenomenon but in fact provokes its self-manifestation to human apperception. That is what great scientific thinking does—it coaxes the phenomenon to appear more fully as what it is. For a phenomenon does not appear where it goes unapprehended. To shine forth in its wonder, the phenomenon needs the thinker every bit as much as the thinker needs the phenomenon.

Schrödinger engages in plenty of explanation in *What Is Life?*, yet what makes it such a thoughtful book is its drive to account for the broader, even metaphysical implications of what he submits to lucid scientific analysis. What are we to make of life's defiance of the laws of physics? Where does our knowledge of life's reproductive mechanisms leave us in understanding our place in the cosmic order of things? And most importantly, who are we who seek to know the nature of things like life and matter, and their interaction? These are the kinds of questions that lurk in the penumbra of Schrödinger' book, for what interests him crucially is to bring scientific explanation to an outer perimeter from which, or beyond which, thinking may draw near the occult nature of reality.

This is most evident in the epilogue to *What Is Life?* Its brief, fragmentary, and altogether bizarre foray into the issue of "Determinism and Free Will" seems to come out of nowhere, and from what I can tell it has been largely ignored by even the most ardent Schrödinger enthusiasts. The epilogue's eccentric afterthought has next to nothing to do with the matter at hand, yet Schrödinger insists that he has earned the right to turn his thinking to the deeper problem of what his investigation into the question of life all means in the end: "As a reward for the serious trouble I have taken to expound the purely scientific aspects of our problem *sine ira et studio*, I beg leave to add my own, necessarily subjective, view of the philosophical implications" (86). Let us briefly review what Schrödinger takes to be the "philosophical implications" of his physics of the living cell.

In less than four pages Schrödinger seeks to resolve the apparent contradiction between the following two premises, each of which (so he claims) is true: "(1) My body functions as a pure mechanism according to the Laws of Nature; (2) Yet I know, by incontrovertible experience, that I am directing its motions" (86–87). Neither of the two premises has been established, by the way, given that: (1) Schrödinger has shown that the body is in fact a very exceptional phenomenon whose laws are far from mechanical; and (2) there has been no discussion in the preceding pages of who or what directs the body's motions. Nevertheless, Schrödinger presumes to reconcile the two premises by peremptorily declaring: "The only possible inference from these two facts is, I think, that I—I in the widest meaning of the word, that is to say, every conscious mind that has ever said or felt 'I'—am the person, if any, who controls the 'motion of the atoms' according to the Laws of Nature" (87).

The statement is not as innocuous as it sounds, since for Schrödinger the one who controls the motion of the atoms is God—or the transcendent force we might call God. Indeed, he does not shrink from declaring that were he to phrase his conclusion in simple words, he would be obliged to declare, "I am God Almighty" (87). "Consider," he writes, "whether the above inference [that I am God] is not the closest that a biologist can get to proving God and immortality at one stroke" (ibid.). He goes on to appeal to the authority of the Upanishads, where wisdom consists in the realization that "the personal self equals the omnipresent, all-comprehending eternal self" (ibid.). He appeals furthermore to the mystics of the Christian tradition, as well as to the spontaneous certainty of "those true lovers who, as they look into each other's eyes, become aware that their thought and their joy are *numerically* one—not merely similar or identical" (ibid.). The main thesis here is the following: "consciousness is never experienced in the plural, only in the singular." Expressed even more decisively, "consciousness is a singular of which the plural is unknown." Schrödinger's conclusion: "there is only *one* thing [in the universe] and what seems to be a plurality is merely a series of different aspects of the same thing, produced by a deception (the Indian MAJA)" (89).

These are extravagant claims indeed, leading one to suspect either that Schrödinger the scientist went on holiday here, or that he had been reading too much Aldous Huxley at the time.[4] The epilogue performs a sudden somersault, which propels us out of the realm of scientific reasoning and into a completely different realm of mystical speculation. In the form it takes here, that mystical somersault is neither great science nor great thinking, yet I believe it would be a mistake to dismiss its seriousness or divorce its intent from Schrödinger's scientific vocation. On the contrary, I see the epilogue's leap as a parting reminder that scientific knowledge of the material world throws us back upon the essential mystery of things, including the mystery of those who pursue science in the first place, namely ourselves.

What Schrödinger discusses so inadequately in his epilogue to *What Is Life?* he treats in a much more serious, rigorous, and systematic vein in his other little book, *Mind and Matter.*[5] I cannot summarize here all the fascinating phenomena—physical and psychical—that Schrödinger deals with in this provocative series of lectures. Suffice it to say that he is primarily concerned here with two "antinomies," as he calls them, which are reminiscent of, yet not identical to, the two "premises" mentioned above. Schrödinger phrases the first one as follows: "all our knowledge about the world around us ... rests entirely on immediate sense perception . . . yet in the picture or model we form of the out-side world, guided by our scientific discoveries, all sensual quali-ties are absent" (153).The second antinomy can be phrased as follows: while the mind is a prime actor in the world, the *place* where mind touches matter is unlocatable, perhaps even non-existent. Let us take these two "antinomies" one at a time.

Discussing the way the world, in our scientific descrip-tion of it, is deprived of sensual qualities, Schrödinger dwells at length on the example of the color yellow. He writes, "If you ask a physicist what is his idea of yellow light, he will tell you that it is transversal electro-magnetic waves of wave-length in the neighborhood of 590 millimicrons" (153).Yet if you ask him where the sensation "yellow" fits into his picture, the physicist will answer that it does not enter his picture at all. All he knows is that "these kinds of vibrations, when they hit the retina of a healthy eye, give the person whose eye it is the sensation of yel-low" (ibid.).The sensation of color *as* color cannot be accounted for, neither by the physicist's description of light waves nor by the physiologist's description of retinas, nerve fibers, brain pro-cesses, and so on. "We may be sure there is no nervous process whose objective description includes the characteristics 'yellow colour' or 'sweet taste', just as little as the objective description of an electro-magnetic wave includes either of these character-istics" (155).

In his ensuing discussion of the sensation of sound, Schrödinger arrives at the same conclusion, namely that while we have an exact physicalistic understanding of the characteris-tics of sound waves, and an equally exact understanding of the physiological mechanisms by which the inner and outer ears register them, "neither the physicist's description, nor that of the physiologist, contains any trait of the sensation of sound" (157). This is a prodigious paradox, for all scientific knowledge passes

through our senses (even when it relies on computers and measuring instruments), yet sensation as such remains alien to science. Why? Because the experience of color, touch, taste, and
sound takes place in a realm that lies outside of the purview of
science. The experience of sound "simply is not contained in
our scientific picture, but *is only in the mind* of the person whose
ear and brain we are speaking of" (158). This mind, it turns out,
is altogether inaccessible to the objectivist reach of science.

Thus Schrödinger's first antinomy blends into the second
antinomy regarding the place where mind and matter intersect.
By "mind" Schrödinger means the "subject of cognizance" as
well as of sentience. The reason why sensation, emotion, and
thought are absent from the scientific picture is because, "without being aware of it and without being rigorously systematic
about it, we exclude the Subject of Cognizance from the domain of nature that we endeavor to understand" (118). By that,
Schrödinger means that "we step with our own person back
into the part of an onlooker who does not belong to the world,
which by this very procedure becomes an objective world"
(ibid.). This stepping back is a necessary condition for scientific
knowledge, which means objective knowledge. Schrödinger is
aware that it is a "high price" to pay: "I continue to regard the
removal of the Subject of Cognizance from the objective world
picture as the high price paid for a fairly satisfactory picture
[of the world]." Yet he often reiterates his conviction that it is
not an altogether *unfair* price to pay. In this regard he is unlike
Carl Jung, for example, who "blames us for paying this ransom."
Schrödinger quotes Jung, who lamented:

> All science [*Wissenschaft*] however is a function of the soul, in
> which all knowledge is rooted. The soul is the greatest of all
> cosmic miracles, it is the *conditio sine qua non* of the world as an
> object. It is exceedingly astonishing that the Western world (apart

from very rare exceptions) seems to have so little appreciation of this being so. The flood of external objects of cognizance has made the subject of all cognizance withdraw to the background, often to apparent non-existence. (quoted in *Mind and Matter* 119–20)

Although at bottom he agrees with Jung on this score, Schrödinger nevertheless cautions that "a rapid withdrawal from the position held for 2,000 years is dangerous" (120). And certainly that caution is well founded. Perhaps what is called for is a *slow* withdrawal. Or even better, an inner transmutation of the objectivist mission of science.

One of Schrödinger's heroes among scientists is Sir Charles Sherrington, an experimental physiologist who embodied two qualities that for Schrödinger are utterly crucial to the practice of science: honesty and sincerity ("the scientist only imposes two things, namely truth and sincerity" [117]). The word "honest" is used repeatedly with respect to Sherrington, about whom Schrödinger writes the following:

> Sherrington, with his superior knowledge of what is actually going on in a living body, is seen struggling with a paradox which *in his candidness and absolute intellectual sincerity* he does not try to hide away or explain away (as many others would have done, nay have done), but he always almost brutally exposes it, knowing very well that this is the only way of driving any problem in science or philosophy nearer its solution. (134)

The paradox in question is the existence of a mind that is the matrix of feeling, perceiving, and thinking but yields no evidence of itself there, where these events presumably occur. Just as Schrödinger had declared in *What Is Life?* that there are not many minds but only one universal mind, so too in *Mind and Matter* he quotes Sherrington to the effect that from the physiological perspective there would appear to be many "sub-minds" in the brain, yet we know that there is finally only one mind. This is the mind that says "I" in the singular, even when it suffers from schizophrenia. This mind has no native home in the brain, according to Sherrington. Here is a long passage from Sherrington's "immortal book" *Man on His Nature*, quoted by Schrödinger:

> Are there thus quasi-independent sub-brains based on the several modalities of sense? In the roof-brain the old "five" senses instead of being merged inextricably in one another and further submerged under mechanisms of higher order are still plain to find, each demarcated in its separate sphere. How far is the mind a collection of quasi-independent perceptual minds integrated psychically in large measure by temporal concurrence of experience? ... When it is a question of "mind" the nervous system does not integrate itself by centralization upon a pontifical cell. Rather it elaborates a millionfold democracy whose each unit is a cell ... the concrete life compounded of sublives reveals, although integrated, its additive nature and declares itself an affair of minute foci of life acting together. ... When however we turn to the mind there is nothing of all this. The single nerve-cell is never a miniature brain. The cellular constitution of the body need not be for any hint of it from "mind." ... A single pontifical brain-cell could not assure to the mental reaction a character more unified, and non-atomic than does the brain-roof's multitudinous sheet of cells. Matter and energy seem granular in structure, and so does "life," but not so the mind. (133–34)

What we encounter here is the puzzlement of a scientist who has looked everywhere within his scientific purview for the mind's connection with matter and has come up empty-handed. It is more than puzzlement; it is astonishment. Sherrington: "Then the impasse meets us. The blank of the 'how' of mind's leverage on matter. The inconsequence staggers us. Is it a misunderstanding?" (122).

Analytic philosophers may well say that it *is* a misunderstanding, and that Schrödinger's and Sherrington's use of the word "mind" is either confused or inconsistent, or both; yet, the analytic philosophers' misery does not touch us here (*la vostra miseria non mi tange*—Beatrice to Virgil in *Inferno* 2). What interests us is the call to thinking in declarations of Schrödinger like the following: "Mind has erected the objective outside world of the natural philosopher out of its own stuff. Mind could not cope with this gigantic task otherwise than by the simplifying device of excluding itself—withdrawing from its conceptual creation" (121). Perhaps it is only by virtue of its abstention from the picture, of its own self-withdrawal, that "Mind" makes room for its "conceptual creation," that is, for the world of space and time as such. Perhaps it is Mind's removal from creation that makes for the movement of the sun and other stars, or the movement of thought that thinks on objects, or the movement of perception that takes cognizance of them through the medium of sensation. Perhaps the world, like a stage or a piazza, needs to be empty if it is to be filled, and that Mind leaves behind, in its wake as it were, this necessary emptiness in whose place the phenomenon first makes its appearance.

Indeed, perhaps it is because Mind has left behind no hard evidence of itself in its creation that reductionists like Edward O. Wilson, Richard Dawkins, Daniel Dennett, and any number of other materialists can claim there is no "ghost in the machine," that mind is an epiphenomenon of the brain and parts of the central nervous system. Or, in Wilson's summation of the reductionist's view: "The human mind is a device for survival and reproduction. . . . The intellect was not constructed to understand atoms or even to understand itself but to promote the survival of human genes. . . . aesthetic judgments and religious beliefs must have arisen by the same mechanistic process."[6] For the reductionist, who by definition is the enemy of wonder, the fact that one cannot find any material evidence for what Schrödinger calls the mind means that the mind is not a trans-organic phenomenon; whereas for Schrödinger, the absence of material evidence in the scientific picture of the brain is the most wondrous of all confirmations of the mind's spiritual and transcendent nature.

When Schrödinger declares that the mind's conceptual creation "does not contain its creator" (121), he is making a statement that cannot be verified, since verification belongs to the domain of the conceptual creation, that is, to the objective world of the natural philosopher. Whether or not we believe, as Schrödinger did, that there are no discrete minds in the plural but only one universal Mind to which all individual minds belong, we can be sure that "the blank of the 'how' of the mind's leverage on matter" is a genuine blank—and not just the result of a linguistic or categorical confusion. We can also trust Schrödinger when he claims the following: "While the stuff from which our world picture is built is yielded exclusively from the sense organs as organs of the mind, so that every man's world picture is and always remains a construct of his mind and cannot be proved to have any other existence, yet the conscious mind itself remains a stranger within that construct, it has no living space, you can spot it nowhere in space" (122). Again, it is not a belief in ghosts that affirms this; it is the "honest and sincere" scientific search for the place where mind and matter intersect—a place that withholds its location even as it constantly "takes place" in animal sensation and human consciousness.

This is what one might call the mystical irony of Schrödinger's thought, namely his awareness that scientific knowledge is a form of *not knowing*, his awareness that there is a blind spot at the heart of the scientific world picture—not just any blind spot but one that opens the objective field of vision itself. As the "subject" of sensation and cognition, the mind is the happening of perception itself, the conversion of matter into meaning, the translation of nature into world, the opening of the "ontological eye" that sees things *as* what they are. This happening of world-disclosure is what science does not see as it scrutinizes and explains what appears within its field of vision.

Schrödinger calls on us not to overcome but to acknowledge the irony of the human condition: to acknowledge it as ineluctable and ultimately fruitful. Such acknowledgment is not easy. It means living with, or better living *in*, the elusive mystery of who we are—and that means living in the mystery of not knowing finally who we are, insofar as we are sentient and thinking beings. This in turn means accepting the fact that selfhood is *not* located in the interior of a person's body. We are so accustomed, Schrödinger says, to locating the conscious personality "inside a person's head—I should say an inch or two behind the midpoint of the eyes," that we forget that this localization "is only symbolic, just an aid for practical use" (123). We may, as Schrödinger writes,

> observe several efferent bundles of pulsating currents, which issue from the brain and through long cellular protrusions (motor nerve fibers), are conducted to certain muscles of the arm, which, as a consequence, tends a hesitating, trembling hand to bid you farewell—for a long, heart-rending separation; at the same time you may find that some other pulsating bundles produce a certain glandular so as to veil the poor sad eyes with a crape of tears. But nowhere along this way from the eye through the central organ to the arm muscles and the tear glands—nowhere, you may be sure, however far physiology advances, will you ever meet the personality, will you ever meet the dire pain, the bewildered worry within this soul. (124)

Finding oneself at such a loss, one is confronted by the mystery of what our scientific picture of the world fails to contain.

By the same token one is confronted by the mystery of what that picture *does* in fact contain, namely the stubborn existence of matter. Nothing is more mysterious finally than matter, in its refusal to yield the secret of its connection to mind. The more we learn about matter in its animate and inanimate modes, the more we fail to grasp its nature. This failure has nothing to do with Bacon's "broken knowledge" and everything to do with the way the world of matter resists humanization—all the more so when it lends itself to scientific explanation. We simply cannot, nor will we ever, recognize ourselves in the picture, short of a mystical vision that takes us beyond our human limitations.

In *What Is Life?* Schrödinger probes the disconnection between the laws of physics and the phenomenon of life. In *Mind and Matter* he probes the disconnection between our lived experience of sentience, cognition, and emotion on the one hand, and the material substrate on which that experience depends, on the other. Scientists are not particularly fond of disconnections, and Schrödinger suggests on various occasions that it takes a great deal of intellectual sincerity to acknowledge the blanks, the impasses, and the dead ends of scientific inquiry. That is one reason he admired Sir Charles Sherrington, who in his "absolute intellectual sincerity" did not "hide away or explain away" the paradoxes with which he struggled but "brutally exposed" them, "knowing very well that this is the only way of driving any problem in science or philosophy nearer its solution" (134).

This is the kind of sincerity we do not always find in the scientific discourse of our times, which is often reluctant to acknowledge how much is left out of its increasingly delimited, objectivist picture of matter, above all of living matter. Certainly when it comes to the relation between mind and matter, there is a widespread tendency to hide or explain away the paradoxes through reductionist or purely materialist schemes rather than to confess the inadequacy of such schemes when it comes to matter's infusion by spirit. We do not ask of science to abandon its drive to explain natural phenomena through natural causes. The lesson one draws from Schrödinger is that one may ask of science that it strive to become more thoughtful, perhaps even more respectful, with regard to the *mysterium* that permeates the world of nature. Instead of becoming more thoughtful, however, science seems determined to become ever more defensive with respect to suggestions that "there are more things between heaven and earth than are dreamt up in your philosophy."

In a *New York Times* editorial, physicist Paul Davies recently declared that a rigid separation between science and faith is untenable, since science has a faith of its own when it comes to believing in the existence and immutability of the laws of nature.[7] "Clearly," writes Davies, "both religion and science are founded on faith—namely on belief in the existence of something outside the universe, like an unexplained God or an unexplained set of physical laws, maybe even a huge ensemble of unseen universes, too. For that reason, both monotheistic religion and orthodox science fail to provide a complete account of physical existence."

Davies relates how over the years he has often asked his physicist colleagues where the laws of the physical universe come from, and why they are the way they are. The replies vary from "that's not a scientific question" to "nobody knows." The favorite reply, he writes, is that "there's no reason why they are what they are—they just are." But Davies insists that at the very least there is a paradox here: on the one hand, we believe that the existence of laws certifies that nature is open to rational investigation and explanation; on the other, we believe that those same laws are without reason. The "we" here includes those scientists who insist that science has no business dealing with such paradoxes, and who feel that any attempt to do so would threaten the integrity of science itself. Davies's article in fact elicited an enormous response in the blogosphere, much of it hostile and much of it from his fellow scientists, who took Davies to task for even suggesting that science and religion are not as far apart as some believe.

Schrödinger—who one must assume is a hero to many of those who attacked Davies—is not nearly as bothered as these detractors by a promiscuous relationship between science and religion. Early in *Mind and Matter* he addresses the paradox of "*one* world crystallizing out of the many minds." Later he discusses the paradox confronted by Sherrington regarding *one mind* arising out of manifold "sub-brains." Schrödinger writes:

> I submit that both paradoxes will be solved . . . by assimilating into our Western build of science the Eastern doctrine of identity. Mind is by its very nature a *singulare tantum*. I should say: the overall number of minds is just one. I venture to call it indestructible, since it has a peculiar timetable, namely mind is always *now*. There is really no before or after for mind. There is only a now that includes memories and expectations. (135)

Schrödinger's speculations about "Mind" in the singular—so reminiscent of Averrois's Universal Intellect—are beyond the pale of science, to be sure; yet, it is science that led him to them. In his words: "But I grant that our language is not adequate to express this, and I also grant, should anyone wish to state it, that I am now talking religion, not science—*a religion, however, not opposed to science, but supported by what disinterested scientific research has brought to the fore*" (135).

One cannot say that in the past half century science has moved in the direction Schrödinger envisioned when he spoke of "assimilating into our Western build of science" other kinds of doctrines—doctrines more spiritually oriented than the militant materialism of the Western models. If anything, Western science has become ever more narrow-minded, if I may be permitted the pun, and ever more dominated by its drive to explain as opposed to reveal. This explanatory drive in turn is bound up with the drive of modern technology to achieve what Descartes called a complete "mastery and possession of nature," even where—perhaps especially where—science and technology wrap themselves in the green robes of environmentalists. These two implacable drives guarantee that much of contemporary science will remain aggressively objectivistic as it continues to advance and to realize the Cartesian dream, which is rapidly becoming a fait accompli.

It is impossible to imagine a universe more astonishing than ours, or a phenomenon more miraculous than a living cell, yet without a sentient mind to take cognizance of it, the world remains a mute, colorless, impalpable, and altogether indifferent place. Schrödinger could not get himself to believe that the world had to wait for a wholly contingent evolutionary development (the animal brain) in order to take cognizance of itself. The animal brain is a "very special contraption" that facilitates the propagation and preservation of certain species. For millions if not billions of years many life forms maintained themselves without such contraptions, and many today still do so. "Only a small fraction of them (if you count by species) have embarked on 'getting themselves a brain'" (135). This scientific fact raises the overwhelming question for Schrödinger. Before certain creatures, and in particular man, acquired brains, was the world a glorious spectacle without witness? "Should it all have been a performance to empty stalls? Nay, may we call a world that nobody contemplates even that [that is, a world]?" (ibid.).

The question is so disturbing that Schrödinger goes on to ask, "But a world existing for many millions of years without any mind being aware of it, contemplating it, is it anything at all?"

(ibid.). In a statement that assures us that he was *not* a dualist, or at least not a conventional dualist, Schrödinger declares that it is a misnomer to say that world is "reflected" in a conscious mind. "The world is given once. Nothing is reflected. The original and the mirror-image are identical. The world extended in space and time is but our representation" (136).

This "romance of the world," as he calls it—the romance that the world had with itself prior to the evolution of the brains that became the substrate for conscious minds—is perhaps the greatest blind spot in the picture of human knowledge. One does not know what to make of it. In Heideggerian terms, we would say that without *Dasein* there is no *Sein*, hence that prior to *Dasein*'s advent in the world there was no world to speak of. Nature existed, to be sure, yet it had no being.[8] Schrödinger does not use Heidegger's language of Being. Instead, without resolving the paradox or dispelling the blindness it condemns us to, he says that sometimes a painter will smuggle into a painting an unobtrusive self-portrait, as Michelangelo did in his Last Judgment fresco. Or a poet will do something similar, as when Homer gives a discrete portrait of himself in the blind bard who sings of the Trojan War in the halls of the Phaikans. Schrödinger declares, "To me this seems to be the best simile of the bewildering double role of the mind. On the one hand mind is the artist who has produced the whole; in the accomplished work, however, it is but an insignificant accessory that might be absent without detracting from the total effect" (137).

Schrödinger's reflections on mind are so unabashedly speculative as to be easily dismissed by scientists and philosophers alike. Yet one need not defend the truth-value of his claims in order to affirm that those reflections show why he was a great thinker. It takes a great scientific thinker to think *through* the picture and beyond it, even if there is nothing that thinker can do to coerce its author into the picture ("I do not find God anywhere in space and time—that is what the honest naturalist tells you. For this he incurs blame from him in whose catechism it is written: God is spirit" [139]). The difficulty or even impossibility of speaking of certain matters does not mean that he who raises his voice about them is a mere dreamer or idle speculator.

Wittgenstein famously ends his *Tractatus* with the proposition, "Whereof one cannot speak, one must remain silent." One lesson we learn from *Mind and Matter* is that the intrinsic limitations of human knowledge, especially in its objectivist manifestations, is *not* an excuse for silence. On the contrary, our refusal to remain silent about that whereof one cannot speak is the best evidence for what we might call the life of the mind. That is why we can go along with Schrödinger when he declares:

> Most painful is the absolute silence of all our scientific investigations towards our questions concerning the meaning and scope of the whole display. The more attentively we watch it, the more aimless and foolish it appears to be. The show that is going on obviously acquires a meaning only with regard to the mind that contemplates it. But what science tells us about this relationship is patently absurd: as if mind had only been produced by that very display that it is now watching and would pass away with it when the sun finally cools down and the earth has been turned into a desert of ice and snow. (138)

It is to be hoped that this absolute silence of scientific research, in its increasingly remorseless objectification of nature, will not turn the earth into such a desert long before the sun finally cools down. It is to be hoped that the stalls will not go empty again before the show itself comes to an end.

# Schrödinger's Trouble

## How Quantum Mechanics Got Created
## with a Logical Loose End

ROBERT B. LAUGHLIN

Let me say at the outset that, in this discourse, I am
opposing not a few special statements of quantum
physics held today (1950s), I am opposing as it were
the whole of it, I am opposing its basic views that
have been shaped 25 years ago, when Max Born
put forward his probability interpretation, which
was accepted by almost everybody.[1]

By the early 1950s, Professor Erwin Schrödinger had become
so exasperated over developments in quantum mechanics, a science he had helped create twenty-five years before,[2] that he
openly disavowed any association with them. His behavior seems
strange, even childish, given the breathtaking advances in atomic
theory,[3-7] chemistry,[8-10] the theory of metals,[11-16] electrodynamics,[17,18] and radioactivity[19-22] that his work had engendered, and
that were still underway at the time. Old people are often like
this, unfortunately. It's scarcely surprising that he was often dismissed as jaded, irrationally in love with his own ideas, and incapable of listening to other people, no matter how reasonable
their arguments.[23]

However, this is not right. For one thing, Schrödinger produced a steady stream of refereed technical papers during this

time on quantum experiment,[24,25] field theory,[26] correlation,[27] superconductivity,[28] cosmology,[29-31] unification,[32] nuclear chemistry,[33] and cosmic rays.[34] None of these was earth-shattering, but none was lunatic raving either. This makes sense, for a person doesn't just go from powerfully persuasive to delusional overnight. He admittedly dabbled in philosophy[35-41] and even poetry,[42] but there's nothing wrong with that, and anyway it doesn't indicate that a person has lost his mind. He certainly had the mathematical wherewithal to understand other people's calculations and appreciate the impressive consistency of these calculations with experiment. The most significant thing, however, is that we are, to this day, still debating the matter that so upset him and ultimately caused him to walk away from the beautiful thing he had created—what quantum mechanics means.

The piece missing from this picture, and which causes it to make perfect sense when reinstated, is Schrödinger's brutal intellectual mugging back in 1927. This sordid bit of history continues to be shrouded in confusion, for the maneuver made by Schrödinger's opponents was highly effective. Yet, the problem is actually quite easy to see once you know where to look: Schrödinger's mathematical description of matter, the one we use today for everything from chemistry to subnuclear particles, is deterministic. It is an equation meaning that a moment from now a quantity will equal what it is right now plus a small correction that you can calculate exactly. The quantity is an abstraction called a wave function, but this is a technical detail. The important thing is that you may mathematically repeat the correction process ad infinitum to exactly predict the situation at all times in the future. But quantum mechanics itself isn't deterministic, at least according to the Copenhagen interpretation.[43] Therefore, Schrödinger's wave equation is incompatible with quantum mechanics.

One of the key bits of evidence that this strange inconsistency is a problem of academic power rather than of science is the schizophrenic way professional physicists continue to deal with it. For example, you would be hard-pressed to find even a single person who would claim that Schrödinger's equation was incompatible with quantum mechanics. In fact, most would say that Schrödinger's equation *is* quantum mechanics. However, you can easily find people who will tell you spooky stories of wave function collapse, that is, behavior not predicted by Schrödinger's equation but nonetheless important for measurement. These ideas are mutually exclusive, of course. If Schrödinger's equation doesn't predict wave function collapse, and you find this collapse anyway in an experiment, then the equation must be wrong sometimes—which is to say, it must be wrong. Having exhausted your patience with wave function collapse, this same person might then switch to lecturing you on atomic clocks,[44] a highly deterministic technology based fundamentally on quantum mechanics and described with enormous accuracy by Schrödinger's equation.

It is also significant that openly questioning Copenhagen is taboo—a singular and very curious exception to the usual scientific ethic that everything can and should be questioned. People foolish enough to raise the issue are instantly labeled crackpots. Nonetheless, essentially everybody who does quantum calculations for a living knows that wave functions don't collapse. They won't say this openly, because it will land them in hot water if they do, but get them alone after work with some alcohol in them and you'll hear, along with other private confidences, "You know, wave functions don't really collapse." The significance here, of course, is that you don't bother making genuinely self-evident things dogmas. It isn't taboo, for example, to question whether dropped objects fall downward. It's just too obvious that they do.

Like many intellectual sleights of hand, this one is difficult to spot because it's based on a half truth. Max Born's original observations were quite correct.[45] Simple Schrödinger descriptions of physical situations do indeed routinely fail to agree with experiments unless the square of Schrödinger's wave function is interpreted as a probability—in which case the agreement becomes perfect. However, this doesn't mean that measurement comes down like a lightning bolt from heaven, doing something terrible and nondeterministic to the system not described by the Schrödinger equation. It does mean that the experiment plus detector form a humongous quantum-mechanical system which obeys the Schrödinger equation faithfully during the measurement process and which generates uncertainty the same way billiard balls on a table do, through chaotic instability.[46] In other words, the probabilistic behavior is a consequence of the Schrödinger equation, not an exception to the Schrödinger equation. It wasn't possible to put the argument the right way around in 1927 for the simple reason that the electronic computers necessary for solving the many-particle Schrödinger equation hadn't been invented yet. Thus the Copenhagen interpretation of quantum mechanics is, for all practical purposes, just a functional reversal of the roles of cause and effect in the theory that quantum measurement facilitated by inadequate computational means.

While it's highly unlikely that Heisenberg and Bohr consciously set out to deceive anybody, it's also clear how they might subconsciously have liked the reversal. Academics then as now struggled constantly over appointments and money, and used such power as they had to promote themselves and their students, thereby accumulating even more power, thereby promoting themselves even more, and so on. You can just imagine then the panic that must have ensued when some unknown professor from Zürich actually accomplished what these gentlemen claimed they had accomplished but actually hadn't. Heisenberg in particular was in a sticky wicket, as one might say, for he had published a "complete theory" of quantum mechanics the year before[47] at the tender age of twenty-three, and was rumored to be on his way to Stockholm. Now he could only watch in horror as his abstruse algebra got swept away by the elegance of some truly high-class theoretical physics generated by an older man—work that naturally turned out to be mathematically equivalent to his own in every way.[48] Bohr's problems were almost as serious. Notwithstanding the Nobel Prize for which he was so famous,[49] he faced the consignment of his celebrated atomic model to the dustbin of history, the dissipation of his authority, and the ruin of his young assistant Heisenberg, on whom he had staked everything. Not surprisingly, Heisenberg had a few uncharitable words to say about Schrödinger, and vice versa. In a 1926 letter to Wolfgang Pauli he wrote[50]

> The more I ponder the physical part of Schrödinger's theory, the more abhorrent I find it. One should imagine a rotating electron with charge distributed over the entire space and axes in a fourth and fifth dimension. What Schrödinger writes about the visualizability of his theory . . . I find it crap.[51]

Schrödinger remarked similarly at a colloquium in Zürich:

> Now the damned Göttingen people are using my beautiful wave mechanics to compute their shitty little matrix elements.[52]

Whatever the reason, the Copenhagen interpretation affected damage control with utterly dazzling brilliance. Whether Bohr and Heisenberg deliberately set out to marginalize Schrödinger we may never know, but their success was total. Without affecting any experimental prediction or appearing in any way to be surly, they effectively preempted his work by making it a footnote to their own rather than the other way around, as should rightly have been the case. Thus when they announced at the Fifth Solvay Meeting in 1927[53] that

> We regard quantum mechanics as a complete theory for which the fundamental physical and mathematical hypotheses are no longer susceptible to modification

they were effectively saying that Schrödinger's theory was very beautiful and centrally important to quantum mechanics but, sadly, not always correct, and that anybody who valued it above the correct theory had no respect for experiment and was therefore an incompetent scientist.

The truly ingenious aspect of undermining Schrödinger's work in this way was that it couldn't be contested scientifically. The reason is simply that interpreting a thing isn't a scientific concept but a political one, as in, "I interpret this vote to mean that people want fiscal responsibility." Experiments can't and don't in general challenge interpretations unless they're very startling, and even then often fail when the old interpretations are too convenient. That's why ideas in science often get entrenched merely by being first—particularly if the experiments in question are chronically confused. The experiments in 1927 were certainly chronically confused. For example, it was accepted practice to define 'measurement' as activity conducted with an apparatus so large that its quantum properties couldn't be discerned. Indeed, the moment an apparatus got small enough to manifest its quantum character clearly, making it possible to tell whether or not things were described by Schrödinger's equation without further postulates, the apparatus was declared to be part of the experiment and disqualified as a means of addressing the question! The situation has improved somewhat now with the advent of nanoscience and its ambition to shrink experimental instruments to the size of proteins. A case in point is the recent report of quantum-mechanical diffraction of buckyballs.[54]

The strategy wouldn't have worked, of course, had the path from the deterministic Schrödinger equation to nondeterministic quantum interference been obvious. However, it isn't obvious. When you first observe particles impacting, one at a time and randomly, on a scintillation screen but slowly building up a grainy image of a double-slit pattern, you are impressed. It appears to be some kind of magic. However, the reason you're impressed, as we now understand, is that the waves in question propagate in configuration space.[55] That means wave amplitudes must be assigned to every possible arrangement of the particles, not just to every possible position in space, as would be the case with a sound or light wave. There are several simple demonstrations of this fact—for example, the slight spectroscopic shifts among the isotopes of hydrogen, or the sharpness of atomic spectra generally—but it's so counterintuitive and difficult to believe in the abstract that doubt wasn't completely dispelled until the late 1950s and 1960s, around the time of Schrödinger's death, when the matter was attacked by computer. However, the basic idea had been worked out in 1926, ironically by Heisenberg.[56]

The strategy also wouldn't have worked if people hadn't found its mysticism so useful. Notwithstanding the popular image of theoretical physics as philosophy, it's actually more like a trade guild[57] and, like all other trade guilds, has secret knowledge by which its members make their living. It couldn't be any other way, actually, for if scientists didn't have special abilities that other people lacked, governments wouldn't need to pay them to do research! The teaching mission of universities efficiently obscures this problem, in that professors paid to teach must deny that they're secretive, sometimes even to themselves, because teaching is the exact opposite of keeping secrets. However, their denial belies a cold economic reality that nicely explains, for example, why very good researchers tend to be very problematical teachers, and vice versa. Thus, the fact that Copenhagen didn't make any sense at all was actually good, in that it enabled you to appear transparent while actually being extremely opaque. You could openly report the results of your work in scientific journals, meanwhile saving back a key part of the computational technology for transmission to graduate students (apprentices) and post-docs (journeymen) by word of mouth as "intuitive skill," thus maintaining the enterprise. And as time progressed, there appeared the happy, unexpected bonus that the mysticism per se was extremely popular with the undergraduates.

Unfortunately, Schrödinger, like Einstein (with whom he got along extremely well), was not a team player.[58] He was an outsider who had not come up through the system by the usual channels, had no powerful supporters in physics or political chits to cash, and worst of all, didn't understand the great economic value of being obscure. His idea of theoretical physics was that you just understood a thing and explained it clearly to people. In retrospect, it's fortunate for all of us that he was so naïve, for he effectively saved quantum mechanics, transforming it from a nightmare of confusing mathematics to something genuinely elegant. But it wasn't so fortunate for him. Not understanding exactly whom he was tangling with, or what was likely to happen if he presented very powerful and creative people a fait accompli that effectively destroyed their livelihoods, he walked bravely into the lion's den and promptly got devoured.

Nine years later, evidently in complete despair over what had happened, he published "The Present State of Quantum Mechanics," in which he introduced his famous cat.[59] His objective in doing this was to repudiate Copenhagen by showing that one of its implications was manifestly absurd. He imagined a box containing a radioactive nucleus, a Geiger counter, an actuator, a cyanide pill, a bucket of acid, and a cat. Were the nucleus to decay, the Geiger counter would detect the decay product and send a signal to the actuator, which would then dump the pill into the acid, killing the cat. However, real radioactive decay is quantum-mechanical, so you must think about the whole box as a giant quantum system. Solving the equations for this system, you find that it evolves into a combination of two states, one with the nucleus decayed and the cat dead, the other with the nucleus intact and the cat alive. Copenhagen then says that a "measurement" is required to determine which of the two states you actually have. This implies that looking into the box kills the cat.

It didn't work, of course. The damage couldn't be undone. Rather than repudiating Copenhagen, as it should have done, Schrödinger's cat became a poster child of Copenhagen, a symbol of transcendent weirdness that students all over the world to this day feel obligated to master if they want to be taken seriously—rather like the insights of Lao Tsu.[60] It appears in the titles of a recent popular book on quantum physics,[61] excellent scientific papers on quantum optics[62] and superconducting electronics,[63] a well-known article on quantum computing,[64] and several fine science fiction pieces.[65,66] And, in a final terrible twist of fate, Schrödinger has come to be remembered primarily as the inventor of this cat (with its reversed meaning) rather than as the inventor of wave mechanics.

Thus, as a consequence of an intense and complicated human drama that played out in 1926, quantum mechanics entered this world with a strange metaphysical deformity which can't be easily explained, doesn't seem to have any experimental implications, and won't yield to reason. It has bedeviled generations of physicists, each of which has found its own way around the blockage—once it figured out what the blockage was—although in a fashion that respected the Copenhagen taboo, thus sparing the next generation no grief. Richard Feynman created a beautiful graphical technique for attacking quantum electrodynamics that people still interpret as different from traditional quantum mechanics but is, in fact, simply a restatement of Schrödinger's equation.[67] Phil Anderson identified a quantum organizational phenomenon called broken symmetry as the real issue underneath Schrödinger's cat.[68] Lev Gorkov developed a Schrödinger description of scattering in metals that completely evaded the Born probability hypothesis.[69,70] Even Murray Gell-Mann went so far as to say in his 1976 Nobel acceptance speech[71] that

> Niels Bohr brainwashed a whole generation of physicists into believing that the problem [of the interpretation of quantum mechanics] had been solved fifty years ago.

But no respectable scientist works openly on the meaning of quantum mechanics because, well, it isn't science.

Given this miserable state of affairs, it's not at all surprising that Schrödinger produced relatively few papers on quantum physics after 1933, that his reputation as a thinker became strangely tarnished, or that he wound up deeply alienated. Despite being constantly lauded and referenced by other experts in the field and doted on in textbooks, he must have been persona non grata among the people who mattered, for having tipped the applecart. The fact that he had been right just made things worse. Moreover, he remained extremely dangerous all his life, for he was a genuine scholar, a proven revolutionary, and a man with the most powerful incentive imaginable to overturn Copenhagen—that it was a monstrous injustice done to him personally.

Unfortunately, Schrödinger's difficulty is something fairly common in academic life, albeit in less egregious forms. Professors in virtually any discipline will ruefully recount how a strangely misty set of ideas has blocked progress, yet still requires obeisance from everyone who wants to be a serious player. It's something we have to live with, like death and taxes. However, there can be no doubt that it is a terrible and shameful blight on everything universities stand for. The morally right response when you encounter it is not acquiescence and sober collegiality but utter outrage—like Schrödinger's. Yet the words you use should not be mousy, deferential things like "oppose" but mighty ones such as those of Shakespeare:[72]

> Blow, winds, blow, and crack your cheeks! rage! blow!
> You cataracts and hurricanes, spout
> Till you have drenched our steeples, drowned the cocks!
> You sulpherous and thought-executing fires,
> Vaunt-couriers to oak-cleaving thunderbolts,
> Singe my white head! And thou, all-shaking thunder,
> Smite flat the thick rotundity o' the world!
> Crack nature's moulds, all germens spill at one
> That make ungrateful man!

# Exorcizing Schrödinger's Ghost

*Reflections on 'What Is Life?' and*
*Its Surprising Relevance to Cancer Biology*

MICHAEL R. HENDRICKSON

No other disease has mobilized societal resources the way cancer has. Thousands of researchers have spent billions of dollars worldwide on cancer research. The lion's share of this has gone to tens of thousands of studies on cancer genetics—a field thriving since the 1980s—the results of which flood not only the medical journals but also the popular media. Almost daily one hears of the discovery of some new 'cancer gene' or some pathbreaking research result that is going to lead to a cure for some type of cancer. And yet, the abundance of scientific discoveries has not translated into cures. While there has been spectacular improvement in the treatment of a collection of relatively uncommon cancers (childhood malignancies come to mind), mortality rates for most adult types of cancer are either stable or increasing, particularly as more long-term survival data become available. Moreover, whatever success there has been is due not to advances in 'gene therapy' but either to technological improvements that facilitate early cancer detection, or to refinements in the conventional, nongenetic, generic modes of therapy: surgery, radiation, and chemotherapy ("cut," "burn," and

"poison"). The results of the efforts in the field of cancer genetics, while significant, have yet to produce the long prophesied
"breakthrough."

Equipped with powerful genomic methodologies, cancer
researchers in the twenty-first-century post–Human Genome
Project era hold out the promise of 'personalized medicine' and
'targeted gene therapy'. Indeed, the very decision to fund the
sequencing of the human genome was largely driven by the expected medical benefits that would follow. But will genomics
work—or are we in for more disappointment? Why haven't we
been able to cash in our growing understanding of cancer at
the molecular genetic level for more dramatic improvements in
treatment?

These are the ultimate questions I address in this essay. However, they can be addressed only indirectly: by elucidating the
fundamental assumptions that underlie the research project of
cancer genetics, by historicizing the conceptual framework that
made that project possible, and by interrogating its origins. From
this examination, a strange narrative emerges, which has Erwin
Schrödinger at its center. Schrödinger plays a paradoxical role
in this narrative: his views can be held responsible for both the
enthusiastic emergence of molecular biology (specifically, cancer
genetics) and, indirectly, for this radical reconceptualization. A
single book, Schrödinger's *What Is Life?*, not only provided, as
I will show, the conceptual foundations for the euphoric development of molecular biology but also anticipated the latter's
ultimate explanatory inadequacy, and pointed the way toward a
meaningful alternative. Schrödinger was both constructor and
deconstructor.

Cancer genetics can be understood only as part of the larger
scientific and conceptual landscape shaped by what I will call the
Schrödingerian Perspective (SP). The mainstream view of cancer

that is still prevalent today flourished in the last quarter of the twentieth century. This view has it that cancer is fundamentally a genetic disease caused by a small number of *oncogenes*, mutated versions of the normal genes involved in cell replication and cell death. This view of cancer seemed to promise that knowing which genes are defective would pave the way for highly specific genetic interventions 'targeting' precisely those genes. Thus, the oncogene research project is based on two interrelated assumptions: first, that genes (and their material instantiation, DNA) play the primary role in 'programming' or 'coding' the normal activities of the cell; second, that dysfunctional genes are primarily responsible for disease (including cancer). The primacy that these assumptions accord to genes is best captured by the ideological construct of DNA as the 'Master Molecule': genes (DNA) 'direct' all life processes by providing the 'instructions' for the development and functioning of organisms.

'Coding', 'programming', 'directing': the powerful metaphors that have become indispensible to our understanding of genetic material can be traced back to Schrödinger's *What Is Life?* In his 1943 Trinity lectures, ten years before James Watson and Francis Crick elucidated the structure of DNA, Schrödinger proposed a model for the material basis of the abstract 'gene' inferred by classical geneticists. He proposed an *'aperiodic crystal'* that had the structural stability (hence 'crystal') and the combinatorial richness (hence 'aperiodic') to contain the blueprint required for the development and functioning of living forms. Schrödinger, in effect, molecularized the classical gene. Additionally, he posited a *'code-script'* that mediated the extraction of the instructions in the genetic material to direct the organism's development and function. He amplified the genocentrism of classical genetics—roughly, the view that the sole agency for development and physiology resides in the genetic material—with the metaphors that attributed to this aperiodic crystal the roles of both "architect's plan and builder's craft," making it thus responsible for the organized complexity of the cell (and, in multicellular life forms, the entire organism).

The influence of Schrödinger's aperiodic crystal and codescript cannot be overestimated. These notions, forming one of the two most important clusters of ideas contained in *What Is Life?*, foreshadowed and inspired the well-known explosive development of molecular genetics in the following decades: notably, the double-helical DNA of Watson and Crick, the discovery of the 'genetic code', and the formulation of the theses that came to be known as the 'Central Dogma' of molecular biology, which proclaims the unidirectional flow of 'information' from DNA to RNA to protein. Thus, the propositions of *What Is Life?*—from which the metaphors are inseparable—shaped a larger conceptual landscape, the Schrödingerian Perspective, which was *preformationist* (it held that the 'plan' or 'program' for the organism was preformed, coiled up in the DNA, the "ghost in the chromosomes"); *genocentric* (it located agency exclusively in the genes); and *reductionist* (it held the view that "we are our genes").

The thirty or so years that followed were characterized by an aggressive pursuit of the Schrödingerian program. Crucially, the extensive toolkits of genetic engineering were developed during that time. However, the overwhelming success of the program contained the seeds of its deconstruction. The flood of data it produced paradoxically exposed its limitations. By providing an unvarnished glimpse into the complexity of living systems, this data forced the researchers to attend to the importance of context, connectivity, emergence, distributed causality, and self-organization in the biomolecular world. The complexity of living systems challenged the notion of an all-powerful Master Molecule; the data exposed the explanatory inadequacy of the SP and generated the demand for an alternative conceptual framework.

This alternative, which I will call the Post-Schrödingerian Perspective, or for reasons that will later become clear, the Systems Biology Perspective, replaced preformationism with epigenesis, the view that the organism is created anew with each reproductive cycle; it is characterized not by genocentrism but by the notion that causal agency is distributed over the entire cell (or organism); and instead of reductionism (the notion that the whole is completely explainable by an analysis of its decontextualized parts), the Post-Schrödingerian Perspective (PSP) emphasizes the necessity of considering context, emergent properties, and the importance of the relationship of parts in constructing the whole. The PSP shifts the emphasis from individual genes to the workings of the genome as a whole. Moreover, the genome is increasingly being viewed as only one of many developmental and physiological resources available to the cell. Agency for both the development of the organism and its ongoing function is distributed over the entire cell and, for multicellular living forms, over the entire organism.

So, the construct of a Master Molecule, with its notions of linear causality and centralized agency, was now beginning to yield to other images: the spontaneously generated order of an eddy in a stream or, again, the emergent, intelligent behavior of an ant colony; organization without an organizer. Offering an alternative to the Schrödingerian Perspective, systems biology draws on two core concepts: networks and self-organizing systems. Networks were familiar to the pre-Schrödingerian biochemists; systems biology rediscovered them and deployed them more broadly. However, for the notion of self-organization, the PSP had to look outside biology to non-equilibrium thermody-

namics, the province of physicists and chemists. These researchers, in turn, had drawn their inspiration from a second cluster
of ideas contained in the very same book that had inspired the
Master Molecule perspective: Schrödinger's *What Is Life?* This
second cluster concerned the thermodynamics of living forms
(bioenergetics). Given that a conspicuous property of living
forms is their high degree of cellular organization, Schrödinger
puzzled over how this organization is possible in a universe in
which the Second Law of Thermodynamics mandates increasing
disorder. Schrödinger's suggestions for how order might arise and
persist were elaborated, over the following fifty years, into the
thermodynamics of open systems (open, that is, to the flow of
energy and matter) maintained far from equilibrium (Schneider
and Kay 1995; Schneider and Sagan 2005). This is a quite general
theory that governs both living and nonliving systems in which
order is spontaneously generated. Examples of physical, lifeless,
self-organizing systems include the heat convection patterns and
the elaborate patterns that arise in a variety of chemical systems.
These 'dissipative systems' are, under specific reproducible circumstances, capable of spontaneously generating order without
the need of any 'architect'.

Thus Schrödinger's second set of ideas in *What Is Life?* provided an important element of the conceptual framework—self-
organization—necessary to rethink the now troubled genocentric program his first set of ideas inspired. The traces of both of
Schrödinger's ideas are evident in twenty-first-century biology.

This shift in metaphors has gradually penetrated cancer biology, again because of a flood of molecular data generated in
this field that made the local version of the Master Molecule
story seem less and less plausible. For example, an embarrassing
plethora of 'cancer genes', currently reckoned in the hundreds,
continues to appear; and worse, it would seem, each individual

cancer features its own unique complement of 'cancer genes'. More and more, cancer is being thought of as a dysfunction of networks that includes, but is not limited to, genetic elements. Cancer, then, is a dysfunction of the cancer cell taken as a whole, not one or more cancer genes. Beyond the complexity of the cancer cell, an individual cancer exhibits developmental and evolutionary complexity.

In this essay, I have two parallel stories to tell: one story recounts the shifts from the SP to the PSP in molecular biology; the other story recounts the corresponding shifts in cancer biology. After sketching Schrödinger's contributions in the first section, I will in the second section follow the fate of Schrödinger's primary (and amazingly prescient) theses about heredity: the 'aperiodic crystal' and the 'code-script' in the 'Heroic Age' of molecular biology. I begin the cancer story by detailing the oncogene theory. In the third section I describe the complexity crisis in both molecular and cancer biology. The resolution of this crisis involved a recognition of the importance of self-organization. The original home of the concepts of self-organization (organization without an organizer) and emergence was thermodynamics. To understand this we need to turn to Schrödinger's concerns about bioenergetics and developments in non-equilibrium thermodynamics of dissipative systems, which I cover in the fourth section. This prepares the way for the concluding section, which sketches a twenty-first-century view of molecular and cancer biology. In this section, I will return to Schrödinger's suggestion that 'new laws' might be required to convincingly answer the question posed by his book's title *What Is Life?*—laws he speculated would be as radically novel as the quantum mechanics and the theory of relativity of his own time. I'll argue that while no 'new laws' have been required, what *was* required was the shift in perspective (SP à PSP) that I've detailed.

I conclude this section by briefly indicating some contemporary, nonmainstream attempts to provide an unvarnished answer to Schrödinger's original question, "What is life?" In the final section, I draw some morals from the Schrödinger tale.

## Schrödinger's Contributions to Molecular Biology[1]

Schrödinger, one of the founders of quantum mechanics, was instrumental in the creation of molecular genetics, and thus played a central role in radically transforming the conceptual framework of biology. He was influential in three ways.

First, his wave formulation of quantum mechanics, for which he received the Nobel Prize in physics, was essential to Linus Pauling's theory of the chemical bond. As Pauling observed:

> The development of molecular biology has resulted almost entirely from the introduction of the new ideas into chemistry that were stimulated by quantum mechanics. It is accordingly justified, in my opinion, to say that Schrödinger, by formulating his wave equation, is basically responsible for modern biology. (Pauling 1987: 228)

Pauling's theory led to the morphologizing of chemistry; the transformation of the printed symbol for a chemical compound (say, $H_2O$) into a three dimensional structure. This in turn led to structural biochemistry with all the now familiar images of 'lock-and-key' biochemical interactions, such as antibody-antigen binding and the three dimensional structure of proteins, including enzymes with their 'active sites'—the clefts and grooves in which their catalytic properties were exercised. Most importantly, the elucidation of the Watson-Crick three dimensional double-helical model of DNA was an exercise in model-building made possible by Pauling's structural chemistry.

Second, Schrödinger's 1944 book *What Is Life?* anticipated the central successes of the Heroic Age of molecular biology (1953–1970s). These include the informational molecule DNA (his 'aperiodic crystal'), the genetic code (his 'code-script'), and by some accounts, the current framework for bioenergetics, non-equilibrium thermodynamics of dissipative systems (his 'negentropy'). He was the first to use the term 'code' in connection with biology and is widely credited (perhaps erroneously) for the proliferation of information-age metaphors in molecular genetics ('code', 'program', 'transcription', 'translation', 'editing', and so on).[2]

Third, Schrödinger's book motivated career changes for a number of researchers who in subsequent decades became leaders in the new field of molecular biology (Judson 1996; Olby 1971, 1994; Murphy and O'Neill 1995).

## The Schrödingerian World: The Triumph of the Master Molecule

### *Before Molecular Biology: The State of Affairs in 1943*[3]

To appreciate Schrödinger's contributions we must first speak about the classical geneticists' conception of the 'gene', a term first coined in the early 1900s. It is difficult to imagine a world without the double helix and genetic engineering. Schrödinger in 1943 inhabited such a world; the subsequent years are well within the remembered experience of many people living at the time of this writing.

*Classical Genetics.* The quantitative study of genetics began with Gregor Mendel in 1865, working with garden pea plants in his monastery garden. He was the first person to treat heredity (what we would now call transmission genetics) in the modern, scientific, quantitative spirit. The genetic principle for Mendel was a theoretical, nonobservable construct designed to model the transgenerational appearance of traits. Mendel's work remained unknown to the general scientific community until it was discovered forty-five years later.

The terms 'gene', 'genotype', and 'phenotype' were introduced by the Danish botanist Wilhelm Johannsen in 1906. 'Gene' denoted Mendel's discrete genetic 'principles'; 'genotype', the aggregate of those principles; and 'phenotype', the ensemble of observable traits that taken together constitute an organism's observable features.

It was Thomas Hunt Morgan's work that matured into classical genetics. Studying the genetics of the fruit fly (*Drosophila*), Morgan in 1910 elaborated on Mendel's approach and identified genetic factors associated with a wide variety of traits. Morgan employed the same strategy as Mendel: he decomposed the overall appearance (the phenotype) of an organism into a collection of traits and studied the transmission of alternative versions of these traits from generation to generation. Thus, for classical genetics the gene was an *abstraction* inferred from an observable trait. I will call this concept of 'gene' the 'trait gene', or $G_{Trait}$.[4] In the classical geneticist's view, the organism's $G_{Trait}$'s constitute in aggregate the genotype; the genotype is inferred from an examination of the phenotype. The $G_{Trait}$ is a theoretical placeholder for whatever is passed on to ensure the reappearance of the trait in the next generation, and it is this sense of 'gene' that figures in the ubiquitous expression "the gene for X."

So the 'classical gene', $G_{Trait}$, was a theoretical abstraction. The research program of classical genetics could proceed apace in complete indifference to the material nature of the gene. In his 1933 Nobel Prize acceptance speech, Morgan noted:

There is not consensus of opinion amongst geneticists as to what genes are—whether they are real or purely fictitious—because at the level at which genetic experiments lie, it does not make the slightest difference whether the gene is a hypothetical unit, or whether the gene is a material particle. (quoted in Falk 1986: 148)

*Genes Are Located on Chromosomes.* It was known from the light microscopic studies of the behavior of chromosomes in cell division (Boveri and others) that the genes studied by the geneticists (what their material instantiation might be) tended to aggregate into groups, and that whatever the material basis of the gene was, the number of these groups correlated with the number of chromosomes. It was also known that the genetic determinants derived from the two parents in a mating were mingled in the germ cells of the offspring by a process of *recombination*, which allowed for the formation of chromosome composites mixing segments of the two parental chromosomes. This phenomenon was extensively exploited to create a functional map of the relative locations on each of *Drosophila*'s four chromosomes of $G_{Trait}$'s, 'genes for' various characters. Again, all of this research moved ahead without any specific knowledge of the chemical structure of the gene.

The classical geneticists noted that *mutants*, that is, new variants of traits, arise spontaneously, and they attributed this variation to an alteration of gene structure. In the 1920s they discovered that they could produce such mutations by subjecting organisms to radiation; however, they did not understand the mechanism of mutation itself.

In 1927, the Nobel laureate geneticist Hermann Joseph Muller (who worked on radiation mutagenesis in the fungus *Neurospora*) set out two functional properties that any material candidate for the classical abstract gene would have to possess: *'autocatalysis'* and *'heterocatalysis.'* Autocatalysis meant that the material basis of heredity must be faithfully duplicated from generation to generation (up to rare mutations). The heterocatalytic capabilities of genes linked them to the phenotype; somehow the order in the gene was translated into the order of the cell and, more broadly, the organism, as manifest in its phenotype. Moreover, constrained biological variation (providing the variation required for Darwinian selection) must be guaranteed in each generation through a process of *'mutation'.* These mutations are then faithfully passed on to the next generation employing the genetic material's autocatalytic function. Mutation is presumably due to small 'errors' in the replicative autocatalytic process, errors whose frequency can, in the laboratory, be increased by artificial agents such as radiation—Muller's favorite method. This was the state of genetics prior to 1943.

*Biochemistry.* What was known about biochemistry (the chemistry of life's molecules) in 1943? By this time, biochemistry had matured into a separate discipline distinct, on the one hand, from organic chemistry and, on the other, separate from medicine and agriculture. The discipline was committed to a materialist explanation of life that while not violating any chemical or physical laws, instantiated them in ways that required an understanding of biological phenomena, constraints, and functional organization. Expanding on this theme, the biochemist J. B. S. Haldane saw life as a pattern of chemical self-regulating processes that beget similar processes. He compared living things to a flame: the flame retains its structure throughout its existence but, unlike life, does not regulate itself. His use of the flame metaphor for cellular metabolic activity implied a non-equilibrium process in an open system capable of reproduction but also, beyond the limit of the metaphor, self-regulation (Weber 2008).

Much of biochemistry's focus was on metabolism and biosynthesis and on the proteins—enzymes—that made that possible. It was recognized that *proteins*, the largest fraction (after water) of the cell, are the most versatile biomolecules. They serve as structural elements, signal receptors, or transporters that carry specific substances into or out of cells. Protein *enzymes* are catalysts; they allow chemical transformation to occur rapidly at the ambient temperature of the body. Each step in metabolism and synthesis is facilitated by enzymes. It was known that proteins were a polymer composed of twenty kinds of amino acids arranged in a very specific way and connected by covalent chemical bonds.[5] These chains of amino acids (think of beads of twenty types linearly arranged on a string)—*polypeptides*—were subsequently folded into a precise three dimensional configuration required for their function in the cell. Thus, the *polypeptide* is folded into a functioning *protein*. This distinction will be important in later discussion.

Weber (2008) summarizes the state of affairs in biochemistry prior to Schrödinger's writing:

> By the time of World War II it was meaningful to address the question of "what is life?" in molecular terms and fundamental physical laws. It was clear that there were several distinct ways in which matter in living systems behaved in ways different from non-living systems. How could genetic information be instantiated at a molecular level given that ensembles of atoms or molecules behaved statistically? How could biological systems generate and maintain their internal order in the face of the imperative of the second law of thermodynamics that all natural systems proceed with increasing entropy?

Chromosomes, the bearers of genes, were known to consist of a complex mixture of protein and nucleic acids. Although the ordering of genes on the chromosomes was being discovered, there was no definitive knowledge of which of these two was the material basis of the Mendelian gene. The smart money at the time of the writing of *What Is Life?* was that proteins played this role; this in fact was the view Schrödinger held. However, the results of a series of experiments beginning in the 1920s suggested, to the contrary, that the genetic material was deoxyribonucleic acid, or DNA. Further confirmation that DNA was the hereditary material came from the classic experiments on bacteria by Oswald Avery, Colin MacLeod, and Maclyn McCarty at the Rockefeller Institute. This work was ongoing when Schrödinger delivered his Trinity lectures in 1943.

The early experimental work in bacterial genetics that would ground Crick's Central Dogma about the flow of 'information' in living things (see below) was just getting under way when Schrödinger's book appeared. The '*one-gene-one-protein*' hypothesis, which held that a specific gene expresses itself by determining (in a manner not then known) the structure of a specific protein, was first enunciated roughly around the time of Schrödinger's lectures. He was unaware of the work.

### Schrödinger's Theses (What Is Life?) Concerning Heredity and Gene Expression

*Changing the Question.* Schrödinger's first move (being ever the practical physicist!) was to change the question from "What is life?" to the experimentally more tractable question, "What is the physicochemical basis of heredity?" (for heredity, read "transmission of traits"). With this move he aligned himself with the genocentrism of the classical geneticists and set the agenda for identifying the molecular correlate of the classical geneticist's $G_{Trait}$; let's call this the 'molecular gene', or $G_{Mol}$.

Equipped with this more tractable problem, Schrödinger set out two subsidiary research programs for biology. The first focused on heredity (which for him included both reproduction and development) and matured into the early molecular genetics of the Heroic Age. This program was concerned with the transmission of *'order to order'*: first, the order of the hereditary determinants from generation to generation (the problem of 'gene replication', Muller's autocatalysis); and second, the transmission of the order in the genome into the organized complexity of the developing organism (the problem of 'gene expression', a version of Muller's heterocatalysis).

*Theses Regarding Heredity.* In the then prevailing spirit of genocentrism, Schrödinger offered two arresting hypotheses: the 'aperiodic crystal' and the 'code-script'. He speaks first of a simple crystal, say quartz, and comments on its stability but also its repetitive simple structure, which would be incapable of encoding much information. The aperiodic crystal, in contrast, combined the physicochemical stability of quartz with the combinatorial richness (aperiodicity) that would be required of a hereditary molecule containing the instructions necessary to produce an organism. In effect, he proposed a model that would "molecularize the classical geneticist's gene." His second hypothesis, the code-script, provided a gene-centered solution to the problem of converting the 'order' of the genetic instructions into the organized complexity of the cell. For multicellular organisms this meant, first, the order of the exquisitely choreographed process of development and, second, the maintenance of order throughout the adult organism's life, which accounted for the organism's physiology. This hypothesis also involved, somehow, the detection of worn-out molecular parts and their repair.

Listen to what Schrödinger says in 1944 about the role of the genetic material:

> In calling the structure of the chromosome fibers a code-script we mean that the all-penetrating mind, once conceived by Laplace, to which every causal connection lay immediately open,

could tell from their structure whether the egg would develop,
under suitable conditions, into a black cock or into a speck-
led hen, into a fly or a maize plant, a rhododendron, a beetle, a
mouse or a woman. . . . But the term code-script is, of course,
too narrow. The chromosome structures are at the same time instrumental in bringing about the development they foreshadow.
They are law-code and executive power—or, to use another
simile, they are architect's plan and builder's craft—in one.
(Schrödinger 1992: 56–57)

For Schrödinger, the gene is "law-code and executive power"—
"architect's plan and builder's craft—in one."

Here are the first inklings of the '*genome-as-program*' metaphor, a metaphor in which 'code' is to be understood *both* as the
encryption of information and the computer science sense of a
set of instructions (a code that encrypts, legislates, and executes).
François Jacob and Jacques Monod, Nobel laureates for their
work on genetic control mechanisms in the 1960s, amplify this
genocentric thought: "the genome contains not only a series of
blue-prints, but a coordinated program of protein synthesis and
the means of controlling its execution" (1961). This combination
makes Schrödinger's molecular gene an unbelievably powerful
agent. In short, his conceptual stance was highly genocentric; it
was reductionist; and it was preformationist.

*Thesis Regarding Bioenergetics.* Schrödinger's second suggested
research program (discussed later in this essay) matured into
twenty-first-century bioenergetics, the thermodynamics of living systems.[6] He picks up on the pre–World War II concerns
of the biochemist. Here the question was how order was possible at all in a universe governed by the Second Law of Thermodynamics. This law mandates increasing disorder (increasing
*entropy*) in thermodynamically closed systems, a disorder that
will, ultimately, culminate in the heat death of the universe. Yet,

as Darwin famously observed, organisms become increasingly complex during evolution. How are these two observations to be reconciled? This then is Schrödinger's problem of '*order from disorder*'.

## The Heroic Age of Molecular Biology: 1953–1970s

The thirty years following the publication of *What Is Life?* are known today as the 'Heroic Age' of molecular biology, as they led to the elucidation of universal genetic mechanisms such as replication, transcription, and translation.[7] Starting with Watson and Crick's double-helical model of DNA in 1953, this 'age' was marked by the great achievements that culminated in the sequencing of the complete genomes of many organisms, including, in 2001, the human genome.

In this section I will show to what great extent the research program of the Heroic Age was inspired and driven by the theses of *What Is Life?* Adopting Schrödinger's arresting hypotheses of the aperiodic crystal and the code-script, molecular biologists during that period 'cracked' the genetic code, detailed the way in which DNA codes for protein, outlined the basics of cell regulatory mechanisms rooted in allosteric control and the operon model of Jacob and Monod, and discovered the molecular anatomy of a large number of proteins by means of x-ray crystallography.

The Heroic Age also adopted Schrödinger's strategic shift from the question of life to the question of the chemical basis of heredity. This first simplification in perspective was now supplemented by another, methodological simplification: all the research during this period used as experimental organisms bacteria (*prokaryotes*[8]) and viruses that prey on them (*bacteriophages*). In contrast to the classical geneticists' experimental organisms (such as garden peas and fruit flies), these organisms have the advantages of a simple microanatomic structure and a relatively simple genetic structure; they also reproduce rapidly.[9] These methodological decisions prompted some observers to have retrospectively relabeled the 'Heroic Age' the 'Age of Precocious Simplicity'.

The Heroic Age also adopted and elaborated the textual-semantic metaphors Schrödinger introduced. The period in which Schrödinger wrote *What Is Life?* saw the birth of information theory (Shannon), cybernetics (Wiener), and the computer theory and technology (Turing, von Neuman). The terminology and concepts flowing out of these fields provided abundant material for the further elaboration of Schrödinger's code-script metaphor (Kay 2000).

The accomplishments of the Heroic Age are well known; I list them here mainly as a foil for their PSP deconstruction.

*The Aperiodic Crystal: DNA and Gene Replication.* Schrödinger's aperiodic crystal is conventionally seen as anticipating the Watson-Crick double-helical model of DNA. Researchers of the Heroic Age attributed to DNA a number of what they thought were essential characteristics:

*1. DNA is a self-replicating molecule.* The basis of DNA (gene) replication is Watson-Crick base-pair complementarity.

*2. DNA's intrinsic chemical structure guarantees its stability and its information-bearing capacity.* The covalent bonds of the DNA molecule's backbone ensure its stability (no crystal required), while the combinatorial richness of a four-character alphabet (A, T, C, G) ensures a sufficient variety of sequences.

*3. DNA is the sole hereditary resource.* DNA is the hereditary material: genes are the important things that are passed on from generation to generation. They provide the instructions for shaping the building blocks of life—the biomolecules—from

the raw materials of small molecules into the genetically speci-
fied form.

4. *The genome is essentially unchanged during the lifetime of the
organism.* DNA is the inert repository of both the information (instructions) required for the organism's development and
maintenance, and the program that executes these instructions;
DNA is both hardware and software at once. DNA stands aloof
and, with the exception of rare mutational events, remains un-
modified from generation to generation. It is like Aristotle's
'unmoved mover'; it directs everything but is itself unchanged
(Delbrück 1971).

*The Code-Script: The Relationship Between Genotype and Phe-
notype.* How is the order of the genome extracted to create the
organized complexity of the cell? The Heroic Age provided a
partial answer to Schrödinger's question of how the order in the
hereditary material is translated into the structural and functional
order of the organism; in short, how genotype is translated into
phenotype. As discussed above, one function of the gene (molecularized as DNA) is to code for the polypeptides that when
folded, become the functioning proteins of the cell. The rich variety of proteins is due to the combinatorial possibilities of an alphabet made up of twenty letters. The cell uses messenger RNA
(mRNA) as a middleman: portions of DNA corresponding to
a gene are *transcribed* from DNA into the very similar nucleic
acid, RNA, using base complementarity. This RNA transcript
is in turn *translated* by a subcellular organelle, the ribosome, into
a polypeptide. Specific nucleotide triplets (for example, GGC)
are translated into specific amino acids (here, glycine); the biochemical 'look-up table' that provides the translation is the 'genetic code'. The molecules that embody this code are a set of
approximately thirty transfer RNAs (tRNA) corresponding to

each of the possible nucleotide triplets; each of these can bind to their appropriate amino acid. It is conventionally believed that Schrödinger's 'code-script' anticipated this 'genetic code'.

*The Central Dogma.* This molecular narrative of gene expression embodies a strong version of genetic reductionism and has come to be known as the Central Dogma of the Heroic Age. The Central Dogma asserts, in short, that the flow of information from DNA to RNA to protein is unidirectional; this assertion has a number of consequences:

1. *Linear, one-way flow of information.* Information passes from DNA to RNA to protein but never in reverse. The DNA to RNA part of the dogma was overturned early in this period; during the Heroic Age an enzyme was discovered that could transcribe RNA back into DNA: reverse transcriptase. In fact, this enzyme is an absolutely indispensable tool for biotechnology. However, extracting the coding information of the amino acid sequence and converting it into a nucleotide sequence is still not known to occur.

2. *DNA's role is to code for functioning proteins.* It was thought that the primary structure of a protein (the specific sequence of amino acids that compose it) spontaneously folds up into the protein's three dimensional functioning form.

3. *The one-gene-one-protein (colinearity) thesis.* A continuous strip of DNA codes for a continuous strip of amino acids.

4. *Genes code for traits.* DNA codes for proteins, and proteins determine cellular traits. This claim asserts the equivalence of the molecular gene $G_{Mol}$ and the trait gene $G_{Trait}$.

5. *The genome (the aggregate of genes) constitutes a program for the organism.* This generalization of the fourth claim is the molecularized version of Schrödinger's genocentrism. DNA unambiguously, coherently, and definitively programs the cell's phenotype.

Taken in aggregate, these claims support the now familiar metaphors for DNA and its powers: 'the blueprint of life', 'the program for the organism', the 'Book of Life', and so on.

*Seeds of Its Own Destruction.* The research program of the Heroic Age yielded a large number of powerful experimental methods; it gave rise to recombinant DNA techniques and the molecular methods that made possible the rapid characterization and experimental manipulation of both nucleic acids (DNA and RNA) and proteins. This remarkable convergence of theory and technology enabled the great achievements of this period, but by facilitating the collection of data, it also precipitated the undermining of the Schrödingerian genocentric perspective (as we will see later).

### View of Cancer During the Heroic Age:
### The Oncogene Theory

*Cancer Research Before the 1970s.*

Biology and Cancer Research have developed together. Invariably, at each stage, the characteristics of the cancer cell have been ascribed to some defect in whatever branch of biology happens at the time to be fashionable and exciting.
—John Cairns (1978)

What was the view of cancer in this exciting Heroic Age of molecular biology? Cancer is the name given to over a hundred different diseases, all of which have one property in common: uncontrolled cell growth. Cancer has been regarded as a cellular disease since the mid-nineteenth century. There were many theories about what went wrong with normal cells to convert them to malignant cells: microorganisms; dys-regulation of metabolism; hormonal imbalance; or environmental factors. At the beginning of the 1980s the two favorite candidates were viruses and mutations caused by a variety of environmental agents. Treatment consisted of surgery and radiotherapy and chemotherapy; the last two exploit the growth characteristics of cancer cells in a generic way. Until the mid-1970s researchers had no technologies for testing theories of cancer at the molecular level. The introduction of recombinant DNA technologies changed all that; cancer could now be studied at the molecular genetic level.

*The Schrödingerian View: The Oncogene Theory.* It was only recently (1976–84) that cancer came to be thought of as a molecular genetic disease. This was the result of a rapid succession of discoveries by Varmus, Bishop, and others—using the new technology—of *oncogenes.* Thus, from the beginning of the 1980s a consensus formed that cancer was due to abnormalities in a small family of genes that were thought to be responsible for transforming a normal cell into a malignant cell, which then went on to produce a clinically detectable tumor. It became apparent that oncogenes were mutated versions of normal genes which were causally implicated in a variety of crucial cell functions regulating cell growth and cell death. Shortly thereafter, other classes of cancer-related genes were identified: *tumor suppressor genes* and *housekeeping genes.* The different dys-regulations produced by these three classes of mutated genes can be thought of in terms of automobile speed control. If cancer is viewed as a problem of failed control of cell division, oncogenes play the role of a stuck accelerator; tumor suppressor genes are analogous to defective brakes; and failed housekeeping genes (by failing to edit copied genetic DNA) lead to larger numbers of mutations, which in particular promote the development of genes in the first two classes. The Schrödingerian paradigm, as played out in cancer biology, had it that there was a handful of genes, normally

involved in regulating essential cell functions, that became dysfunctional through, for example, mutations.

## The Post-Schrödingerian World:
## The Rediscovery of Physiology and the Organism

Now my own suspicion is that the universe is not only
queerer than we suppose, but queerer than we can suppose.
—J. B. S. Haldane (British geneticist 1892–1964)

While sharing a set of evolutionarily conserved core processes fixed in living forms for three billion years, bacteria and
nucleated cells are vastly different. The names *prokaryotic cell*
(lacking a nucleus, as in bacteria) and *eukaryotic cell* (containing
a nucleus, as in all multicellular organisms) mark this important
difference. Bacteria are thousands of times smaller than nucleated cells. Other differences are striking. The nucleated cell's
complex, compartmentalized anatomy contrasts with the homo-
geneous interior of the bacterium. The nucleated cell's DNA is
distributed over several (the number varies by species) structurally and biochemically complex structures, the chromosomes.
The bacterium's chromosome lies free in the bacterium cytoplasm, and both transcription and translation occur simultane-
ously. In nucleated cells, transcription occurs in the nucleus; the
edited transcript is moved to the cytoplasm, where the ribosome
translates the code. So, the development of an internal anatomy
of compartmentalizing membranes, two billion years ago, was
a major jump in the complexity of life and a major evolutionary development. A second major evolutionary advance was the
coming together of nucleated cells to form *multicellular* organisms, about 1.2 billion years ago. It is only when this occurred
that *development* (or *embryology*) became relevant. How does the

organism develop from a single cell having one genome into, for humans, the adult organism containing several hundred different cell types arranged in a precise order? These two evolutionary developments—eukaryosis and multicellularity—are central because they force an entirely new way of thinking about organisms; namely, levels of organization.[10]

I will simply list, with brief comments, some of the major results (in parallel with features in the previous section) in the molecular biology of eukaryotes, findings that brought the field to a state of crisis.

### Replication: DNA, the High-Maintenance Molecule

1. *DNA, unaided, is not self-replicating.* The brilliance of the double-helical model of DNA structure is that it provided a chemical basis for replication: Watson-Crick complementary base-pairing. However, DNA is a relatively inert biomolecule, incapable on its own of much of anything. DNA replication requires a host of proteins involved in assembling nucleotides in the right place at the right time, and others involved in editing, proofreading, and the repair of damaged segments. Other proteins are involved in coordinating DNA replication with the other complex events that occur during cell division. This was well known to the molecular biologists of the Heroic Age but not emphasized in popular accounts of the Master Molecule. As Richard Lewontin has observed, the popular impression of the self-replication of DNA is about as plausible as thinking that the letter you feed into your copy machine makes five copies of itself unaided. Here is our first hint of the importance of context: DNA is involved in a web of relationships, and replication is possible only with the participation of other elements.

2. *The hereditary stability of DNA is an emergent property.* Schrödinger was largely ignorant of the biochemistry of the day.

The stability Schrödinger sought in the crystalline structure of his aperiodic crystal (a stability that would overcome the disruptive effects of thermal motion) was there all along in the chemist's covalent bond. That said, Schrödinger's quantum mechanics (via Pauling) rationalized the stability of the covalent bond. But genetic stability involves more than stable covalent bonds; it requires the copying of the specific nucleotide sequences to be accurate. A variety of proteins are involved in proofreading and correcting the newly synthesized DNA to keep the overall error rate of DNA replication to a minimum. Speaking of DNA as the vehicle of hereditary stability, Fox Keller observes that "the source of genetic stability was not to be found in the structure of a static entity but that stability is itself the product of a dynamic process." Again, we are confronted with a web of relationships.

*3. The genome is not the sole hereditary resource.* It is a mistake to think that only the parental genomes are passed on to the next generation. The new organism (at least among those that sexually reproduce) involves the fusion of two cells: a paternal sperm and a maternal egg. The sperm consists largely of a nucleus with a tail. The egg, in contrast, is a fully equipped nucleated cell having the full complement of organelles and molecular machinery (e.g., mitochondria with their own DNA derived exclusively from the egg), ribosomes, membranes, cytoskeleton, and messenger RNAs. The fertilized egg consists of the now-merged parental genetic contributions (and whatever epigenetic markings they contain) within a complete cell. The fertilization process can be thought of as equipping a preexisting cell (the egg) with a new genome. We could think of this as the 'Sleeping Beauty' model of development: the cell is quickened into activity by the act of sperm penetration. In a sense, this is to recall the obvious. Bacteria split and eggs get fertilized; obviously, the new organism inherits more than a couple of sets of parental genes. The

German pathologist Virchow summed it up: "Cells come from cells." This was a commonplace of embryology, and the need to remind molecular geneticists of it is a measure of the power of the genocentrism of early molecular biology. Again, DNA by itself is incapable of doing anything; it is chemically relatively inert. The egg supplies the hundred or so proteins involved in the DNA-reading machinery. Indeed, the early cell divisions after fertilization proceed without any transcription of DNA. In addition, the egg supplies all its own machinery, and this appears to be pretty species-specific; one can't make DNA do its thing in any old egg, *Jurassic Park* notwithstanding.

4. *The genome (totality of DNA) is structurally dynamic.* The Schrödingerian image of the aloof 'Director of the Cell' is now known to be deeply misleading. In eukaryotic cells the nucleus has more the character of a dynamic rainforest than a hushed library dedicated to housing the 'Book of Life'. The genome, partitioned into the chromosomes, is sheathed in a layer of proteins, all of which is immersed in a sea of noncoding RNAs and mobile genetic elements, including 'jumping genes' (*transposons*[11]). Barbara McClintock, who won the Nobel Prize for her discovery of 'jumping genes' in corn, commented that the nucleus is a

> highly sensitive organ of the cell, monitoring genomic activities and correcting common errors, sensing the unusual and unexpected events, and responding to them, often by restructuring the genome. We know about the components of genomes [transposons] that could be made available for such restructuring. (McClintock 1984)

In bacteria the genome is a similar hotbed of activity. The passage of bits of genome from bacterium to bacterium is commonplace ('horizontal transmission'), and the mutation rate of the bacterial genome can be modulated depending upon environmental conditions. Adverse conditions lead to less effective proofreading, higher mutation rates, and a larger variety of genomes. This increases the chances that a variant will be produced which can successfully meet the challenge of the altered environment (Goldenfeld and Woese 2007; Zimmer 2008).

Moreover, DNA itself undergoes chemical modification; it is 'marked' by the organism in order to make possible different cell types. Only parts of the genome are read at any given time, and this very much depends on the particular type of cell (the cell phenotype). The neurons of the brain, for example, have to access the genome in a different way from that of the cells of the skin; yet, the genome is the same for both types of cells. This differential access is managed through a kind of bookmarking ('epigenetic changes'[12]) that can be passed on to the descendants of neurons or those of skin cells. Importantly, it is the overall state of the cell that determines whether regions of the genome are marked or not. Here is the first hint of '*downward causation*'. What determines the pattern of markings on the cell's genome is the global state of the cell, a higher level of organization than the genome. In summary, the genomic DNA increasingly appears to be more of a database of genetic resources than Schrödinger's all-powerful director. The cardiovascular physiologist Denis Noble (2006) has likened the twenty thousand or so genes in eukaryotic organisms to the pipes of a very large organ that are differentially recruited depending on which 'melody' the cell is playing at a particular time.

## Gene Expression: The Complex Relationship of Genotype to Phenotype

As we have seen, the application of molecular biological techniques to eukaryotic, multicellular organisms led to a radical reconceptualization of the genome; it also completely transformed our understanding of gene expression—the relationship of genotype to phenotype.

1. *The Central Dogma is undermined.* The Central Dogma is misleading in several ways. If it covers *only* the transmission of coding information, then it is noncontroversially true.[13] If, however, the Central Dogma is taken to mean, more broadly, that *all* information flows from DNA to RNA to protein, then it is manifestly false. There is more information than *coding* information; for example, information is resident in the specificity of lock-and-key molecular complementarity. First, the Central Dogma privileges the causal role of DNA over that of RNA and protein. This is a bias that came with pre-Schrödingerian genocentrism and the subsequent identification of DNA with the trait-gene ($G_{Mol} = G_{Trait}$) (Moss 2003). The choice of the DNA bacteriophage viruses (or 'phages') as experimental model organisms reinforced this bias. Phages amount to stripped-down reproductive machines; they are in the business of injecting their DNA into bacteria and driving the unfortunate host's metabolic machinery to make more phages. Bacteria are a different matter. Bacteria are fully integrated organisms, and all of the elements—DNA, RNA, and protein—are woven seamlessly into a set of mutual dependencies, which can be modeled as a complex network. For example, replication, transcription, and translation all require proteins and RNA. Moreover, both proteins (transcription factors) and RNA (noncoding RNAs) mediate the control of these processes. So, the first departure from the Central Dogma is the expansion of roles for the various types of RNA and protein. This is accompanied by a diminished role for DNA. The all-powerful DNA molecule begins to resemble the queen in a colony of ants, a queen that is less like Elizabeth I and more like the hive's 'ovary'.

The second problem with the Central Dogma arises with

the shift from linear causality to circular or network causality (Sattler 1986; Cohen and Atlan 2005). Traditional notions of cause and effect presuppose a linear relationship between events: A causes B causes C. Once A, B, and C are circularly connected (for example, through feedback loops), we have a chicken-and-egg problem; each is at once the cause and the effect of the other. More problems arise with the incorporation of nodes for biomolecules other than protein and nucleic acids (for example, lipids and carbohydrates). Additionally, the cell's microanatomic context is important; what a cell does is highly dependent on the cells around it. Other difficulties arise with alternative splicing (see below).

2. *Strictly speaking, DNA doesn't 'code for' proteins.* Polypeptides (the strings of amino acids coded for by DNA) have to fold in order to be functional proteins. A specific sequence of amino acids implies its three dimensional structure, but in general it would take an unreasonable amount of time for a polypeptide on its own to 'find' the conformation required for protein function. Other proteins, called *chaperones*, facilitate this process. In short, newly synthesized proteins require other proteins for full functioning.

3. *Genes don't 'code for' traits.* Earlier, when discussing classical genetics, I noted that genes don't literally code for traits. Instead, different versions (called *alleles*) of a given gene are causally implicated in producing differences in a particular trait. For example, the 'gene for brown eyes' plays a causal role in the pigmentation of eyes that would otherwise be blue. This involves the production of a protein (an enzyme) that in turn makes possible the production of pigment (melanin), which imparts color to the iris. Moreover, in the production of a trait, genes usually don't act in isolation; they act as ensembles or networks. Traits are also influenced by environmental factors and the statistical

contingencies of 'developmental noise' (see developmental systems theory, below).

It is crucial to understand that 'gene' is used in at least two distinguishable senses. First is the gene of Mendel and Morgan, that is, an abstract theoretical construct that points to an *observable trait* (say, flower color) which can assume different forms (purple flowers, white flowers). The gene can correspondingly take on different forms (alleles) that track variants of the trait (a 'gene for flower color' that can take on values of 'white' or 'purple'). I have used $G_{Trait}$ for this sense. Talking of genes in this sense often uses phrases like 'gene for X' (as in 'gene for blue eyes'). Taken literally, this language is deeply misleading to the extent that it suggests this gene is solely responsible for the trait. In reality, all that can be inferred is that the gene is causally implicated in producing the trait. In general, the entire genome is responsible for the production of the eye that may be variously pigmented. This perhaps is obvious for eye color; even more so when talking about a 'gene for aggression'. Genes are 'difference makers'; differences in a classical gene cause uniform phenotypic differences in particular genetic and environmental contexts (Waters 2007).

In the decades after Schrödinger's book, 'gene' acquired a second sense, denoting a segment (continuous or discontinuous) of DNA that codes for a messenger RNA. Sometimes the term is confined to protein-coding RNA; other times it is extended to include DNA that is transcribed into RNA that is not translated into protein (that is, noncoding RNA). For this molecular version of 'gene' I have used $G_{Mol}$. L. Moss, who originally made these distinctions, uses Gene-P (P for 'preformationist') and Gene-D (D for 'developmental resource') to distinguish these two senses. I will expand on his meaning in my concluding comments (Moss 2003, 2008).[14]

*4. The genome (the aggregate of genes) doesn't 'program' the organism.* Virchow was right: it takes a cell to make a cell (Cohen and Atlan 2005). It is commonly said that the genome is the program for the organism; this is often made vivid by calling the genome the 'Book of Life'. Both are unhelpful metaphors that echo Schrödinger's earlier fantasy about being able to deduce the organism from a read-out of the genome. The modern version of this is Nobel laureate Sydney Brenner's claim that he could 'compute an organism' from knowledge of its genome. If what is required of a program is that it be unambiguous, coherent, and definite, then the genome fails on all these counts. I summarize the important observations:

*a. Exons and introns.* Most polypeptide-specifying DNA sequences in eukaryotic organisms are discontinuous, consisting of *exons* (coding segments) separated by *introns* (noncoding stretches). The exons are spliced together by a complex molecular machine (the *spliceosome*) to fashion the messenger RNA to be translated.

*b. Alternative splicing.* In many cases, different polypeptides (that will form different proteins) derive from the same gene. Such genes undergo 'alternative splicing', a process that involves the stitching together of different exons to yield different proteins from the same gene.[15] The DNA strip is thus ambiguous.

*c. One polypeptide, many genes.* I spoke earlier of the newly synthesized polypeptide's need for chaperone proteins in order to fold up into a functional three dimensional structure. (Again, the protein coded for by a gene may fold up in a variety of ways and perform a variety of functions.) Moreover, proteins undergo a range of posttranslational modifications all mediated by yet other proteins. There is obviously more to a functioning protein than its coding DNA. Further, the alternatives sketched here are determined by the state of development of the cell and its current needs.

*d. A genome's expression depends on its cellular environment.*
What a genome does depends on its environment. The cloning
of Dolly the sheep involved transferring the nucleus of an ud-
der cell into a sheep ovum. The genome reconfigured itself to
produce a sheep rather than milk. A new organism has inherited
the father's sperm and the mother's egg; fertilization initiates a
'conversation' between the new genome and the maternal cytoplasm. Moreover, the organism is embedded in an environment that plays an essential role in development. Finally, there
is 'developmental noise', the unavoidable probabilistic element
of normal development. For example, when a cell of a developing organism divides, there often is an unequal distribution of
a small number of specific regulatory protein molecules, which
in turn leads to differential rates of development of the daughter
cells' lineage. This is why identical twins (clones developing in
the same environment), for example, have different fingerprints.
In summary, the production of the adult organism involves essential inputs from genome, maternal cytoplasm, the environment, and chance.

*e. A genome's expression depends on the organism's developmental
stage.* The same protein can function in different ways during
different developmental stages of the organism.

*f. Knockout surprises.* It is possible to destroy specific genes in
genetically altered animals, for example, 'gene knockout' mice.
The results of these experiments are often surprising: when some
genes that are judged to be crucial are destroyed, or 'knocked
out', the organism is not unduly compromised. This attests to
the robustness of cellular networks in 'working around' a defi-
ciency like a missing gene. In contrast, other genes that seem less
essential turn out to be lethal when knocked out.

*g. Making new genes: the immune system.* The cells of the immune system fashion antibodies and cell receptors (proteins) from

genes that they make by randomly scrambling otherwise unex-
pressed small genetic elements inherited in the germ line. Thus,
the immune system cells manufacture millions of different genes
that are not encoded as such in the genome. Accordingly, the immune system's ability to recognize antigens—a major determinant
of health or disease—is a function that is absent in the germ line.

### Schrödinger's Perspective Problematized

The Schrödingerian Perspective (SP) acknowledges the existence of levels of organization, but it is *reductionist* in believing
that all higher-level properties can be explained by lower-level
properties. There are no emergent properties; causality is all 'up-
ward', from lower organizational levels to higher organizational
levels. At the molecular level the SP is committed to linear
causality and the unidirectional flow of information embodied
in the Central Dogma of molecular biology. DNA is the only
source of information (*genocentrism*).

The Post-Schrödingerian Perspective (PSP), that of systems
biology, differs on all accounts. The PSP takes levels of organization very seriously. New properties emerge at each level of organization, and these can be causally effective both at their own
level and at lower organizational levels ('downward causation').
There is no privileged level of examination or explanation. At
all levels we have networks rather than linear causal chains. In
particular, at the biomolecular level, a network connects DNA,
RNA, proteins, and other biomolecules. In contrast to the SP,
the role of DNA is diminished and that of protein and RNA
correspondingly expanded. There is a distribution of causal
agency over all elements. The large number of molecular machines (some composed of both RNA and protein) involved in
gene replication and gene expression exemplifies this network
entanglement. The usual discourse of cause and effect doesn't

work for networks. For the simplest network, a circle, we have a chicken-and-egg problem; each is jointly cause and effect. The PSP is *holistic* rather than reductionist. It is not genocentric but locates agency at *all* levels of organization, none of which are privileged. It is, in short, *organism-centered*.

What of *preformationism* and self-organization? We will return to this issue after taking up Schrödinger's second project: the thermodynamics of life, or bioenergetics.

## The Crisis in Cancer Biology

The shift between the Schrödingerian and the Post-Schrödingerian perspectives has reverberated through all of applied genomics, including cancer biology. Before we look at the crisis in cancer biology, it will be instructive to take a brief look at the impact of the PSP crisis on biotechnology (and the 'industrial gene') and on medical genetic disease. In both of these domains the initial optimistic appraisals for successful interventions have given way to more sober estimates.

### Biotechnology and the Industrial Gene.[16]

> However, there are growing indications that in many
> cases either single genes do not affect phenotype, or that
> their influence on phenotype does not arise in a simple,
> obvious fashion. For example metabolic engineering
> efforts to enhance the ethanol production rate in yeast by
> overexpression of each enzyme in the glycolysis pathway
> [alcohol-producing metabolic pathway] have produced no
> significant rate increases. Similarly, after extensive efforts to
> clone all of the genes in the penicillin biosynthetic pathway,
> overexpression of these genes in an industrial strain has no
> effect on penicillin production.—James Bailey (1999)

Biotechnology, which was concerned with the production of commercially valuable biomolecules—organic compounds,

antibiotics, hormones, cancer drugs, and so on—had a powerful financial interest in the outcome of the conceptual crisis I
have detailed above. The projection of the SP into the world of
biotechnology produced the 'industrial gene', a gene that in the
words of one observer, "can be defined, owned, tracked, proven
acceptably safe, proven to have uniform effect, sold and recalled"
(Caruso 2007). In other words, this was a gene with the properties required for the classical geneticist's 'bean bag genetics'. As
we have seen, the postgenomics of the twenty-first century has
provided little comfort for this view. Genes, it was now understood, operated in context; the expression of a particular gene
depended crucially on the state of other genes in its network.
There was no single 'gene for' a particular disease or for a phar-
maceutically important product. Indeed, the consequences for
the 'industrial gene' were dire (Wilkins 2007). The late James
Bailey, a metabolic engineer, reflected the growing PSP in the
epigraph above: you can't force yeast to make more alcohol (or
penicillin) by tinkering with the genes directly involved in the
metabolic pathway. In short, intervening in the workings of a
complex nonlinear network in a targeted way did not lead to
a localized change in that targeted metabolic step; it produced
instead a disequilibrium for which the entire network attempted
to compensate in order to restore the network to its previous
state.

*Medical Genetics.*[17] Medical genomics is another area hit by
the 'Turn to Holism'. The poster child of medical genetics in the
Heroic Age was Linus Pauling's 'molecular disease', sickle cell
anemia. Here was a genetic disease whose roots could be traced
from one misplaced nucleotide in the hemoglobin b-gene,
which led to a crucial amino acid substitution in one of the hemoglobin chains, which in turn led to a malformed hemoglobin

(oxygen-carrying) molecule. This in turn produced deformed (sickle-shaped) red blood cells. These sickled cells plugged up blood vessels to yield the clinical findings of anemia (red cells being destroyed faster) and oxygen deprivation of a variety of organs, leading to multiorgan symptoms and findings. This is the straightforward Schrödingerian tale. There are two points to make. First, the importance of context has emerged. Unfortunately, in reality the sickle cell story is vastly more complicated: modifier genes (genes other than the sickle cell gene) could produce a range of clinical appearances in patients with sickle cell disease. Patients might have no symptoms or, alternatively, might be profoundly ill in early childhood, or fall somewhere in between. And context enters in another way: patients with one abnormal gene and one normal gene are more resistant to malaria than people with normal hemoglobin. Thus, carriers for sickle cell anemia enjoy a selective advantage in regions of the world infested with malaria—precisely the areas (like sub-Saharan Africa) where sickle cell anemia is common. This complicated interplay of genetic context (modifier genes) and environment appears to hold for most of the so-called monogenic disorders. The second point is that the gene which codes for the abnormal hemoglobin has been known for many years, as have the pathogenetic pathways leading to clinical symptoms. While many therapeutic advances have been made exploiting facts of pathophysiology, no practical gene therapy is on the horizon for sickle cell anemia.

### The Crisis of the Oncogene Theory.[18]

Maybe the oncogenes of the animal genome will be boiled down to a very small number of distinct types. Maybe the present oncogene jungle is not so impenetrable after all.
—R. A. Weinberg (1982)

Unfortunately, the number of 'cancer genes' is steadily increasing (the count is now into the hundreds and growing), and more disturbing, each individual's tumor appears to be a unique patchwork of such mutations. Thus, every patient's breast cancer is as unique as that patient's finger prints. For example, initial results have been published by the recently launched *Cancer Genome Atlas*, involving an analysis of eleven individual colorectal cancers and eleven breast cancers. The workers concluded that there were approximately one hundred 'cancer genes' in each case (out of a large group of mutations not known to be associated with cancer), and that there was substantial variation from individual cancer to individual cancer within the same anatomic group (colon or breast). This sounds much like the 'too much information' problem that steered molecular biology into the PSP. A new vision of cancer has coalesced around these observations, which I outline in the section after next.

## Schrödinger's Second Problem: Bioenergetics

Up to now, I have been discussing Schrödinger's concerns with heredity and development. In order to move forward the SP→PSP story, I now need to take up the second half of his reflections in *What Is Life?*: his concerns about bioenergetics. We'll see that the conceptual apparatus that was developed to address these problems essentially enriched the conceptual structure of molecular genetics. In particular, I will need two concepts: first, the crucial idea of a network; and second, physicochemical instances of self-organizing systems. These ideas led to a new vision of life as a self-organizing system without having to posit spooky vitalistic forces, thus providing an alternative to the centralized agency of the all-powerful gene. The incorporation of these ideas (along with some others) into the molecular biology of the late twentieth century has led to the creation of *systems biology*. This is the conceptual framework within which the complexity crisis of postgenomic biology is currently being addressed.

Schrödinger's third Trinity lecture set out a paradox: the Second Law of Thermodynamics (entropy always increases in closed systems) mandates the heat death of the universe; yet, as Darwin observed, complexity appears to increase over the course of evolution. How can these seemingly conflicting theses be reconciled? A related problem: how is order maintained once it is established? Schrödinger's suggested answer for the first problem was that there was nothing in the Second Law that prohibited an increase in local order (say, within a bacterium) as long as the entropy debt was paid for elsewhere in the universe (by the bacterium excreting waste products and heat into the outside world). The second problem was more difficult for him to address, particularly since he was woefully ignorant of the biochemistry of his day. As it turned out, the answer to the problem of how order is maintained required a conceptual framework not available in 1943—namely, the non-equilibrium thermodynamics of dissipative systems. The decades after 1943 saw the full development of the science of dissipative systems by Prigogine and colleagues in Brussels (Nicolis and Prigogine 1989; Prigogine 1997).

## Dissipative Systems: The Organisms
### of Physics and Chemistry

A useful entrée into Schrödinger's paradox is provided by an examination of the emergence of order in a class of physical and chemical systems called *dissipative systems*. Their uncanny resemblance to living things has prompted the name 'organisms of physics'. Dissipative systems are thermodynamically open systems that are maintained far from equilibrium by the continuous flow of matter and energy through them. Under the right conditions, such systems can undergo spatial self-organization with a concomitant enhancement of energy throughput across whatever energy gradient exists.[19] They are, in short, gradient-destroying. The picture to have in mind is an eddy in a mountain stream or, more dramatically, a tornado. Like a living organism, a physicochemical dissipative structure, when conditions are right, transforms from a chaotic swarm of randomly moving submicroscopic particles into a visible form that involves the coordinated motions of millions of particles over an extended territory. It will persist indefinitely, so long as the supply of matter and energy lasts. Bénard heat convection cells and Belousov-Zhabotinsky (B-Z) autocatalytic chemical reactions are celebrated, well-studied laboratory examples of such systems. They all exemplify the emergence of form without the intervention of an organizer; that is, they self-organize. Again, the material composition of the structure is always changing while the form persists. All of this occurs without any centralized direction. If the set-up (context) is right and the physical parameters just right, complex structure miraculously emerges. What does all of this have to do with life?

### Metabolism as Envisioned in the Twenty-First Century

Consider a bacterium, say the common gut bacterium *E. coli*, immersed in an environment that includes only glucose and essential minerals. Miraculously, the bacterium converts the 'order' of the glucose molecule into the 'order' of its constituent molecules, including cell wall, cell membrane, enzymes, structural proteins, and nucleic acids. It does so through a network of enzymatic reactions that collectively constitute the complex web of intermediary metabolism which creates, sustains, and regenerates the bacterium. A characteristic of intermediary metabolism is the extraction of energy from the chemical bonds of glucose and its capture in the form of ATP—the universal energy currency. This currency can then be spent on driving chemical reactions involved in synthesizing, among other things, the complex molecules that are characteristic of life: proteins, nucleic acids (DNA, RNA), carbohydrates, and lipids (cell membrane). We trace this ancient bioenergetic trajectory when we eat, grow and repair, excrete and give off heat. Our material composition is constantly changing; the molecules of yesterday's dinner are incorporated in our constitutive molecules, literally from head to toe.

### Life Is a Slow Burn

Let me now forge a connection between two living and nonliving organized dissipative structures: the bacterium and the flame. A candle flame is a dissipative system: it has a relatively constant form (give or take some flickering) but no persistent material basis. Matter and energy flow through it as long as there is wax to burn and oxygen to sustain it. The burning of a candle and the slow burn of cellular metabolism are both instances of *combustion*: oxidation occurs in both with an overall increase in the world's entropy. Both feature a *structure* (the flame and the organism) that persists as long as energy and matter flow through it. Both are *open* systems: an exchange of matter and energy is required to maintain an orderly form in both instances. For candles it is melted wax taken in and soot and heat eliminated; for organisms it is food taken in and waste and heat eliminated. There are, of course, important differences. Cells and candles operate at very different temperatures. High temperatures are required to break the bonds of the wax molecules; enzymes (proteins) bend the bonds of biomolecules so that the snapping or connecting of bonds can occur at body tempera- ture—enzymes are basically catalysts. The energy of the candle is lost as heat and light; the energy produced by metabolism is captured in a usable form, the energy currency molecule ATP. This currency can then be used to build the complex biomol- ecules constitutive of life. The burning of a candle and the slow burn of the organism: it's the contrast between a gasoline fire in your driveway and the regulated combustion of your car engine as you drive to work. Finally, returning to reproductive themes, consider the difference between the spread of a wildfire and the reproduction of living forms. We will consider this difference in more detail below.

### Schrödinger's Views Reinterpreted

We are now in a position to reinterpret Schrödinger's dis- cussion of 'negentropy', the "sucking order from the environ- ment into the organism." The order (or 'negentropy'[20]) of the glucose molecule's covalent bonds is transformed into the order of, say, a part of a DNA molecule in a set of reactions that in- volve capturing the energy of the sugar in the form of ATP and using that energy currency to fuel a coupled reaction that drives the synthesis of the highly ordered (high 'negentropy' content) DNA.

We leave off our treatment of the history of molecular biol- ogy with the field in a conceptual crisis. The world is revealed to be vastly more complicated than envisioned in the Heroic Age. As we shall see, the conceptual framework required to make sense of bioenergetics is essential to making sense of Schröding- er's first set of concerns—heredity and gene expression.

# Are New Laws Required
# to Explain Life?

Toward the end of *What Is Life?* Schrödinger appears to have second thoughts about the extreme materialist, reductionist, gene-centered position that he has sketched out, and he wonders whether 'new laws', consistent with and supplementing existing laws, might be required in order to explain life. He seems, consciously or unconsciously, to be anticipating the twenty-first-century movement in biology toward the reintroduction of the organism into biology; the movement from a gene-centered to an organism-centered biology.

> Of course, the scheme of the hereditary mechanism, as drawn up here, is still rather empty and colorless, even slightly naive. For we have not said what exactly we understand by a property [trait]. It seems neither adequate nor possible to dissect into discrete 'properties' the pattern of an organism which is essentially a unity, a 'whole'. . . . (quoted in von Baeyer 2004)

> [T]here is just one general conclusion to be obtained from it, and that, I confess, was my only motive for writing this book. . . . [I]t emerges that living matter, while not eluding the 'laws of physics' as established up to date, is likely to involve 'other laws of physics' hitherto unknown, which, however, once they have been revealed, will form just as integral a part of this science as the former. (68)

This tantalizing promise of new laws has been inspirational for some workers (for example, G. Stent, M. Delbrück) and judged ludicrous by others (such as M. Perutz). One concern of this essay is to understand, from the perspective at the beginning of the twenty-first century, whether new laws are required.

At the end of the last section we left molecular biology in a state of crisis produced both by too much information about an exponentially growing molecular parts list and by the discovery of the unexpected and complex interrelationships among these parts. The 'beanbag genetics' of the Heroic Age of prokaryotic molecular biology had completely broken down; in general, there was no straightforward relationship between individual traits and the strip of DNA that coded for a gene, between the $G_{Trait}$ and the $G_{Mol}$. Any particular gene was involved in the production of a number of traits (*pleiotropy*), and an ensemble of genes was typically required to produce a given trait (*epistasis*). Moreover, a given trait could be multiply realized; many genetic pathways could lead to a given phenotype. What became clear was that to explain cell-level, organ-level, and organism-level function would require putting the molecular pieces of this Humpty Dumpty back together again in a way that preserved their original topology. All of the pieces seemed to be enmeshed in a dauntingly complex set of intersecting networks. In passing, we should note that the retreat from the SP was, paradoxically, forced by the very success of the reductionist program at its heart.

The reaction to this crisis has taken a number of forms, all of which entail, to some extent, a Holistic Turn, the abandonment of the belief that complex structures and functions could directly, and completely, be explained by the properties of a limited number of gene products. In short, this is the transition from the reductionism of the SP to the holism of PSP.

There have been at least three systematic published responses to this crisis: (1) one form or another of systems biology; (2) developmental systems theory (DST) and its predecessor, dialectical biology; and (3) neo-Kantianism. Let's take a brief look at each of these.

*Systems Biology: The Answer to Information
Overload and Unexpected Connections*[21]

Without an adequate technological advance the pathway of
progress is blocked, and without an adequate guiding vision
there is no pathway, there is no way ahead.—C. R. Woese
(2004)

Woese, a microbiologist, concisely summed up the state of
affairs. The genomic revolution produced vast amounts of data,
but the SP was inadequate to the task of making sense of it. As
I indicated above, the required supplemental conceptual framework emerged from an unexpected quarter, bioenergetics, which
contributed the concepts of the network and self-organization
(Westerhoff and Palsson 2004).

So, systems biology, in varying degrees, represented a substantial retreat from the SP to the PSP; it emphasized the relationships among biologically important molecules and studied
them as elements of complex networks located more broadly in
the framework of physiology, developmental and evolutionary
biology, and ecology.

The systems biologist Marc Kirschner, when pressed for a
definition, sums it up like this:

> systems biology is the study of the behavior of complex biologi-
> cal organization and processes in terms of the molecular constit-
> uents. It is built on molecular biology in its special concern for
> information transfer, on physiology for its special concern with
> adaptive states of the cell and organism, on developmental biol-
> ogy for the importance of defining a succession of physiological
> states in that process, and on evolutionary biology and ecology
> for the appreciation that all aspects of the organism are products
> of selection, a selection we rarely understand on a molecular
> level. Systems biology attempts all of this through quantitative
> measurement, modeling, reconstruction, and theory. (2005: 504)

Two comments contextualize the shift: first, the computational power of the modern computer was essential to getting this program off the ground; second, most work in molecular biology as of this writing is still largely reductionist in spirit. A growing minority of workers is now employing 'systems biology'; indeed, it is not uncommon to find issues of major scientific journals devoted to this expanding field, and many leaders in molecular biology have adopted the PSP (for example, the Nobel laureates Lee Hartwell and Sydney Brenner).

Systems biology is thus becoming more and more mainstream. The same cannot be said of the other two approaches, outlined below; these find favor mostly with philosophers. In general, they sacrifice experimental tractability (remember Schrödinger's simplifying first move: change the question) for scientific realism. We will explore this further in the Conclusion.

*Developmental Systems Theory and Dialectical Biology*

A second substantial step away from the SP began in the late 1970s with the writings of Richard Lewontin and Richard Levins, a population biologist and an ecologist, respectively. These two (with various colleagues since the late 1960s), both working from a Marxist perspective ('dialectical biology'), have launched, in a succession of books and articles, a sustained attack on genocentrism, the nature/nurture distinction, and more generally, the sharp distinction between organism and environment (Lewontin *Human Variation* 1982; Lewontin, Rose, and Kamin 1984; Levins and Lewontin 1985; Lewontin 1993; Lewontin and Levins 2007). Susan Oyama, greatly influenced by their views, wrote the seminal work setting out the perspective of DST, *The Ontogeny of Information* (Oyama 2000; Oyama *Evolution's Eye* 2000; Oyama, Griffiths, and Gray 2001). The DST perspectives

TABLE 1.  Developmental Systems Perspectives

---

*Joint determination by multiple causes.* Every trait is produced by the interaction of many developmental resources.
The gene/environment dichotomy is only one of many
ways to divide up these interactants.

*Context sensitivity and contingency.* The significance of any
one cause is contingent on the state of the rest of the system.

*Extended inheritance.* An organism inherits a wide range of
resources that interact to construct that organism's life cycle.

*Development as construction.* Neither traits nor representations of traits are transmitted to offspring. Instead, traits are
made—reconstructed—in development.

*Distributed control.* No one type of interactant controls
development.

*Evolution as construction.* Evolution is not a matter of organisms or populations being molded by their environments
but of organismal–environmental systems changing over
time.

---

*Sources*: Oyama, Griffiths, and Gray 2001; Griffiths 2002.

are set out in Table 1; they emphasize distributed causality in
both development and evolution and specifically deny the primacy of one developmental resource (say, genes) over another
(say, proteins).

Developmental systems theory is *organism-centered*. The ap-
propriateness of the metaphor of 'unfolding', that is, the playing out of a preformed genetic program, is denied. Nature (the
preformed program) is inextricably mingled with nurture (environment), rendering untenable the traditional nature/nurture
dichotomy. In place of the metaphor of the 'genetic program' we
have a relationship between phenotype, genotype, and environ-

ment known by the technical term '*norm of reaction*', or NoR. To this mix of genotype and environment must be added 'developmental noise'.[22] In addition, the distinction between organism and environment is blurred. If you consider the termite mound or the beaver dam, it is clear that the organism isn't responding to a fixed environment; it actively constructs its local environment. DST emphasizes *self-organization* (*epigenesis*) in that it holds that the life cycle of each organism is created anew with each generation by ensuring that roughly the same initial conditions are repeatedly reassembled. The new life cycle emerges from these initial conditions; the organism self-organizes. In the earlier section on metabolism we viewed living organisms as autocatalytic dissipative systems; in the DST perspective, they are seen as self-organizing developmental systems. The approach is *holistic* in that there is no primary or privileged locus of causality. Causality is distributed over both genetic (DNA, RNA) and nongenetic elements (proteins, cell membrane templates, and so on). This perspective is integrated into Kirschner's definition above of systems biology (Gerhart and Kirschner 2007; Kirschner, Gerhart, and Mitchison 2000; Kirschner and Gerhart 2006).

### Neo-Kantianism: Systems Closed to Efficient Causation[23]

In 1790 Kant put his finger on the essential difference between the living and nonliving:

> In a machine the parts exist for each other but not by each other; they work together to accomplish the machine's purpose but their operation has nothing to do with building the machine. It is quite otherwise with organisms, whose parts not only work together but also produce the organism and all its parts. Each part is at once cause and effect, a means and an end. In consequence, while a machine implies a machine maker, an organism is a self-organizing entity. Unlike machines, which reflect their maker's intentions, organisms are "natural purposes." (Harold 2001: 220)

When Schrödinger changed the question from "What is life?" to the experimentally more tractable question, "What is the physicochemical basis of heredity?" he was himself aware that he might be answering a substantially different question. There are hints that his worry was well founded. Outside the narrow context of research focused on the origin of life from abiotic precursors and research on artificial life, molecular biologists are largely unconcerned with Schrödinger's question "What is life?" Cornish-Bowden and colleagues quote Jacob (1970) on this point ("Today we no longer study Life in our laboratories"), as well as H. Atlan and C. Bousquet (1994) ("Today, a molecular biologist has no need, so far as his work is concerned, for the word 'life'"). They conclude that "it would be difficult to argue that we understand life any better than Schrödinger did 60 years ago" (Cornish-Bowden et al. 2007). This brings us to the most radical reformulation of the original question, "What is life?" undertaken by two research groups who, unconstrained by tractability issues, take an unvarnished look at that question. They return to Kant's reflections on the nature of life, set out in his 1790 *Critique of Judgment*. These groups' answers are the *(M, R)-systems* of R. Rosen, and the '*autopoietic systems*' of H. Maturana and F. Varela.

Since Descartes, living organisms had been understood in terms of machines. There was, however, an acknowledged fundamental difference between machines and organisms: it took a watchmaker to make a watch, and a jeweler to repair it. In contrast, organisms both made and repaired themselves. Rosen extended this Kantian line of thought in a project captured by the slogan, "A material system is an organism if, and only if, it is closed to efficient causation." Asking, as did Kant, how an organism could be different from a machine, Rosen used the machinery of abstract mathematics (category theory) to work out the details of his thesis. And he came up with a formal description that would accomplish his goal: the 'metabolism and repair system' or (M, R)-systems. Similarly, the Chilean workers Maturana and Varela, following Kant, thought of living organisms as self-fabricating (autopoietic) systems in the sense of both reproduction and ongoing repair and regeneration (Maturana and Varela 1987, 1991; Weber and Varela 2002). There were points in common between their autopoietic approach and Rosen's approach, but autopoiesis put the primary emphasis on the structural organization of organisms and the necessity to enclose them within membranes, whereas Rosen was more concerned with logical organization in terms of formal mathematics. Both approaches emphasize the importance of organizational closure but in slightly different ways. Neither, to date, has had any influence on molecular biology. I include them to make a point about the price of experimental tractability: the danger of missing the original target because you've moved it, but believing you've hit it because you forgot that you moved it.

Schrödinger raised the possibility of 'new laws'. Were any required? It is clear that all three of the outlooks presented here differ sharply from Schrödinger's perspective. Keller asks the provocative question, How does a living organism (an example of organized complexity) differ, on the one hand, from a tornado (emergent complexity) and, on the other, from a computer (another example of organized complexity)? Kant supplied the start of an answer: Watches (and computers) can't make more of their own kind; watches always require a watchmaker. Organisms differ from tornadoes in that rather than being "one-shot, order-for-free" kinds of self-organization, they are the product of iterative processes of self-organization that occur in heterogeneous systems over time (Keller 2007). Returning to Schrödinger's original worry, I think it is safe to say that while no 'new laws' (comparable to quantum mechanics and relativity) have been required yet, a major shift in conceptual orientation has.[24] I have argued that both places in the conceptual landscape—the SP and the PSP—have been repeatedly visited over the years. The difference is that these places were visited with different tools, and some of them allow for deeper insights into the question, "What is life?"

## Conceptualizing Cancer in the Twenty-First Century[25]

Do these perspectives offer anything to the cancer biologist? I think they do. As discussed above, the individual cancer cell inherits the complexity of the normal cell. In particular, cancer genes act in concert to produce their growth-disturbing effects. The methodologies of systems biology are required to make sense of the large number of disturbed genetic networks that characterize individual cancers. But individual cancers are complex in other ways: they are developmentally and evolutionarily complex. First, individual cancers are evolutionary processes (Merlo et al. 2006). Cancer cells are genetically unstable and accumulate mutations as their constituent cells divide. Some of these mutations endow cancer cells with additional properties (such as the ability to spread to the bloodstream and lymph channels—that is, to metastasize). Thus, a clinically detectable cancer (say, more than 1.0 cm in diameter, roughly one billion cells, or thirty cell generations) consists of a large number of subclones of malignant cells, each genetically different from its sisters. The individual cancer begins as a monoclonal growth but rapidly diversifies into dozens of daughter subclones genealogically related to the original clone. There does not appear to be a single, specific, fixed stepwise progression from normal cell to malignant cell in most types of adult cancer. Second, individual cancer subclones are microecological systems. Much of the volume of a clinically detected cancer is composed of non-neo-plastic (benign) cells; these include blood vessels, immune cells, and fibroblasts (benign cells that form the scaffolding or matrix inhabited by cancer cells). Each subclone of cancer cells interacts with non-neoplastic cells to create its own subclone microenvironment; cancer cells thus participate in a microecology. For example, cancer cells secrete factors that lure blood vessels into their neighborhood. Drugs directed at molecules responsible for vessel growth target these saboteurs. In light of these observations, cancer can be regarded as an abnormal developmental system and evaluated within the framework of DST.

Equipped with these further characterizations of cancer, we can now begin to understand the difficulty in its cure. Cancer therapy (radiotherapy or chemotherapy) can be viewed as the imposition of a Darwinian selective force on the genetically and phenotypically heterogeneous collection of subclones comprised in the clinically evident tumor. An analogy may be helpful. Notoriously, antibiotic treatment of many hospital-acquired infections leads to antibiotic resistance. The antibiotic may kill the vast majority of bacteria, but there remain a small number of bacteria that because of genotypic differences are resistant. A cancer's acquired resistance to radiotherapy or chemotherapy can be understood in the same terms; the vast majority of the cells composing a cancer may succumb to a therapy, leaving microscopic collections of resistant cells. These, unfortunately, regrow to make a 'cancer recurrence'.

What can we conclude from this? David Reiff, Susan Sontag's son, writing about his mother's long battle with cancer, reflects on the state of twenty-first-century cancer treatment in a recent *New York Times* article. He seeks the opinions of several active cancer researchers and collects some surprisingly (in light of the popular-press presentation of cancer successes) pessimistic impressions. For example, Mark Greene, whose lab did much of the fundamental work on Herceptin, an important 'targeted therapy' for breast cancer, observes that the best way to deal with cancer is to "treat early, because basic understanding of advanced cancer is almost nonexistent, and people with advanced cancer do little better now than they did 20 years ago" (Reiff 2005).

## Conclusions

What lessons can we extract from the Schrödinger story?

1. *Conceptual change in biology.* I have made the case that molecular biology's conceptual evolution has followed a particular trajectory: a movement from preformationism, reductionism, and genocentrism to a growing appreciation for the role of self-organization (epigenesis), emergent properties, context, and the interaction among multiple levels of the organism's organization. It would be a mistake, though, to see this as a new development in the history of biology. Here is J. Maienschein, a historian of biology, on one aspect of this shift:

> Epigenesis and preformation are two persistent ways of describing and seeking to explain the development of individual organic form. Does every individual start from material that is unformed, and the form emerges only gradually, over time? Or does the individual start in some already preformed, or predelineated, or predetermined way? . . . The debate has persisted since ancient times, and today plays out as genetic determinists appeal to the already 'formed' through genetic inheritance [the twentieth-century genetic programs], while others insist on the efficacy of environmental plasticity [such as developmental systems theory].
> (2005)

She continues by locating this polarity in the larger history of ideas:

> In 1899, American biologist William Morton Wheeler suggested
> that there are just two different kinds of thinkers. Some see
> change and process, while others see stability. Heraclitus,
> Aristotle, physiology, and epigenesis characterize one way of
> looking at the world, while Parmenides, Plato, morphology, and
> preformationism characterizes another. These are, Wheeler felt,
> stable and persistent classes . . . just the nature and details of their
> differences have changed over time. (2005)

Aristotle, the first published embryologist, coined the term *epi-genesis*; he was impressed, as he watched chicks develop from eggs, with the creation of form anew with each generation. This position became increasingly untenable at the dawn of modern science in the sixteenth and seventeenth centuries. Matter in motion seemed incapable of producing form, on its own, without the introduction of some form of vitalism. Preformationism was one answer to this problem. A familiar seventeenth-century image of preformationism is the homunculus coiled up in the sperm; the twentieth-century version was foreshadowed by Schrödinger's "architect's plan and builder's craft," the blueprint of life, the "ghost in the chromosomes." Systems biology and, most radically, developmental systems theory, return us to the Aristotelian idea of emergent form undirected by a Master Molecule.[26] Developmental systems theory emphasizes a variety of nongenetic (epigenetic) factors that are coresponsible (with genomic elements) for the construction and maintenance of the organism. In place of an earlier time's required vitalism, now discredited, non-equilibrium thermodynamics has given us self-organization.[27] So, no new laws seem to be required to explain life, merely a return along a well-worn trajectory but, crucially, with the spectacular methodologies of genetic engineering. If there *are* new laws of the fundamental sort suggested by Schrödinger, they will probably be those that govern biological networks.

Moss sees the two senses of 'gene' tracking on the preforma-
tionist-epigenetic polarity:

> The preformationist gene (Gene-P) predicts phenotypes but only
> on an instrumental basis where immediate medical or economic
> benefits can be had. The gene of epigenesis (Gene-D), by con-
> trast, is a developmental resource that provides possible templates
> for RNA and protein synthesis but has in itself no determinate
> relationship to organismal phenotypes. The seemingly prevalent
> idea that genes constitute information for traits (and blueprints
> for organisms) is based, I argue, on an unwarranted conflation of
> these two meanings which is, in effect, held together by rhetori-
> cal glue. (2003: xiv)

*2. Necessity of simplification and its pitfalls.* As we have seen,
the Heroic Age of molecular biology was made possible by virtue of two simplifications: first, a change in the original question, "What is life?" to the experimentally more tractable, "What
is the chemical basis of heredity?"; second, a change in the experimental organism from the classical geneticists' eukaryotic
organisms—Mendel's garden peas and Morgan's fruit flies—to
structurally much simpler, stripped-down organisms—*E. coli* and
Delbrück's bacteriophages. These two simplifications made possible the accomplishments of the Heroic Age of molecular biology and yielded, first, a collection of basic universal biomolecular
mechanisms that underlie gene replication and gene expression
(base-pair complementarity, genetic code) and, second, a powerful set of techniques (gene cloning, gene sequencing, polymerase
chain reactions) that constitute molecular engineering. As we
look back over the history of molecular biology, these were absolutely essential prerequisites for understanding, at a molecular
level, the complexity of multicellular organisms.

These simplifications do however carry with them the
danger of premature generalization. Recall Monod's comment:

"What's true for E. coli is true for the elephant." While this claim is true for many of the core biochemical processes, it turned out to be spectacularly incomplete. It took the application of bioengineering techniques to the molecular and cell biology of eukaryotic organisms to make this obvious.

*3. Role of metaphor in science: necessary yet potentially dangerous.* As A. Rosenbleuth and N. Wiener have written, "The price of metaphor is eternal vigilance" (quoted by Lewontin 2001). Models and metaphors are indispensable for the practicing scientist's work.[28] First, metaphors in science function as *heuristics*, devices that help scientists think about theoretical objects by relating them to familiar objects. The behavior of billiard balls is taken as a model for the behavior of gases; or a well worked-out theory serves as the guide for the construction of a new theory: Schrödinger appealed to classical wave mechanics to motivate his formulation of quantum wave mechanics. Metaphors also play a *conceptual* role in formal theory. Models and metaphors that are at one time highly fruitful can, over time, outlive their usefulness. The Master Molecule and the associated Schrödingerian Perspective is an instance of this, which we have explored; as we have seen, it has been replaced by a variety of epigenetic metaphors in the twenty-first century.

But the use of metaphors also carries risks. One is reification, the passage from, for example, "the organism is *like* a machine" to "the organism *is* a machine."[29]

Metaphors serve as *rhetorical* devices; scientists in their role as educators or in their efforts to obtain funding for their research employ these routinely. As a number of commentators have emphasized, many of the metaphors that scientists use to explain biological concepts are at best misleading, and at worst simply inappropriate.[30] We have examined some of these: the 'self-replicating molecule', the 'genetic program', the 'Book of Life', the

'Master Molecule'. Additionally, metaphors form connections with other metaphors, often causing unfortunate consequences. For example, the blueprint metaphor suggests that one's genetic constitution is solely determinative of one's appearance or behaviors:

> We used to think our fate was in the stars. Now we know, in large measure, our fate is in our genes. (James Watson, *Time*, Mar. 20, 1989)

Following close on the heels of these genocentric metaphors is (too frequently) the associated thought that one can't change the consequences of one's genetic constitution. Here is one rebuttal:

> We cannot think of any significant human social behavior that is built into our genes in such a way that it cannot be shaped by social conditions. (Lewontin, Rose, and Kamin 1984)

Genes are undeniably an essential developmental resource, but they are only one of a number of factors that jointly determine phenotype and behavior.

4. *Public and professional discourses about genes and their powers: the theater of genetics.* In reading Schrödinger's *What Is Life?* I had the feeling that this extraordinary book is at once as modern as the latest *New York Times* article proclaiming the discovery of the 'gene for X', and as outdated as an early 1960s issue of the journal *Nature* or *Science*. This points to the fact that the public discourse about genes and their powers (especially as they relate to cancer) is radically different from contemporary professional discourse. The public discourse reflects an updated version of Schrödinger's 1944 genocentric outlook; it is haunted by the 'Ghost of Schrödinger'. We find in the popular press locutions like "the genome is the program for the organism" or "DNA replicates itself," or statements proclaiming the existence of a "gene for X" or of "selfish genes."

Why is there this disconnection? Part of the explanation is that the alternative perspective of systems biology has only recently made inroads into the professional literature. Indeed, most mainstream molecular biology articles remain Schrödingerian. But I think that there are also deeper reasons.

A wide range of factors influences the public discourse about genes; this is discussed in a huge literature.[31] I'll mention only a few. The public typically requires instruction to make sense of the mind-numbing complexity and arcane terminology of molecular genetics. This allows popularizers (and proselytizers) substantial scope in the kinds of metaphors they employ. The resulting blend of metaphors that permeate the public discourse is complex and is influenced by factors other than education. For example, the metaphor of the 'genome as program' was used extensively to sell the public on the multibillion-dollar Human Genome Project (Kevles and Hood 1992). Human disease had already become geneticized (and later molecular-geneticized) when the funding debate was underway. The sequencing of the genome offered the promise of exposing an individual's defective 'program' and making it available for therapeutic 'debugging'. Any metaphor that strayed very far from the 'Master Molecule' and its promise of relatively easy therapies would be an obstacle to fundraising. Another pressure for pushing the Master Molecule metaphor was the biotechnology boom in the 1980s, a program whose appeal depended on localized points of intervention to fashion profitable commercial molecules.

5. *Tenacity of metaphors of localized over those of distributed control: exorcizing the Ghost of Schrödinger.* What accounts for the abiding popularity of the organizer as an explanation of form? One notion is that if there is organization, there must be an organizer; organization doesn't just emerge. Another way of formulating this is 'top-down' versus 'bottom-up' control. Lewontin

and his colleagues argue historically: 'top-down' thinking is a legacy of sixteenth-century physics. Forces are applied to insensate matter; matter doesn't have the capacity to generate global organization. That constructs the playing field. They argue that the 'bottom-up' generation of order would be a more natural way of thinking for the medieval mind; it's when mind became modern that emergent order became unimaginable. It took the rediscovery of natural examples of self-organization—tornadoes and eddies—to reimagine alternative sources of order.

    6. *What is life, after all?* Schrödinger's *What Is Life?* played two important roles in the history of molecular biology. His views about hereditary transmission informed the Heroic Age of molecular biology, while the elaboration of his views on the thermodynamics of living systems laid the groundwork for non-equilibrium thermodynamics and the twenty-first-century demonstration of the possibility of order without an organizer. Can we find his traces in modern definitions of life?

    In 2001 the microbiologist F. M. Harold offered the following definition of life:

> "Living organisms are autopoietic systems: self-constructing, self-maintaining, energy-transducing autocatalytic entities" in which information needed to construct the next generation of organisms is stabilized in nucleic acids that replicate within the context of whole cells and work with other developmental resources during the life-cycles of organisms, but they are also "systems capable of evolving by variation and natural selection: self-reproducing entities, whose forms and functions are adapted to their environment and reflect the composition and history of an ecosystem." (Harold 2001: 232, quoted by Weber 2008)

Such a twenty-first-century perspective represents a fulfillment of the dual insights of Schrödinger in 1944.

    7. *Reverberations in applied genetics: why the war on cancer is*

*so difficult to win.* I began this essay by expressing some worries about the future of cancer therapy. As we have seen, the shift in perspective that I have presented suggests that the usual cancers (for example, breast, colon, lung, prostate) are going to be difficult to cure. As many have observed, by the time a cancer is advanced it is spectacularly complex along several dimensions, and it notoriously resists both conventional and early attempts at 'targeted' therapy. The simple linear causality of the Central Dogma embodied in the Schrödingerian conceptual scheme held out the hope that once the genes responsible for cancer were discovered, a localized 'debugging' of the defective 'program' would effect a cure. We now see that the Schrödingerian Perspective is inadequate to deal with an individually unique process that is both genetically and epigenetically complex. Unfortunately, for the foreseeable future, successes in the war on cancer will continue to be measured by incremental advances in prevention and in early diagnosis, which is amenable to conventional nontargeted therapies.

*Translated by Lisa Ann Villareal*

# Keeping the Singular, Risking Openness

*Erwin Schrödinger's Way of World Experience*

HANS ULRICH GUMBRECHT

Erwin Schrödinger had a profound distaste for speaking and writing about himself, especially in public situations. An impressive collection of records and evidence makes this clear beyond any doubt, even if his contemporaries were otherwise divided as to whether Schrödinger should be looked on as particularly modest or as arrogant to the point of hubris.[1] For example, the biographical texts archived in 1928 on the occasion of his induction into the Austrian Academy of Science were not penned by Schrödinger himself. A year later, he began his inaugural address before the Prussian Academy of Science with the sentence, "Let me begin by acquitting myself as quickly as possible of the awkward duty imposed on all academic inaugural lectures, namely of speaking about myself"—and then limited himself to commending his science professors at the University of Vienna at the turn of the century. He felt little more duty-bound to deliver an autobiography on the occasion of his acceptance of the Nobel Prize for Physics in 1933, and the few interviews granted by Schrödinger provide sparse insight into his personality.[2] Not until 1960, in the midst of his seventy-third year, after surviving an acute phase of inflammation of the lungs from which he would never fully recover, and a mere year before his death in January 1961, did Erwin Schrödinger decide to tackle the "embarrassment of autobiography," as he called it. What reasons he may have had, indeed whether he pursued any particular purpose at all in writing *My Life, My Worldview*,[3] will remain forever obscure. We know only that he reread multiple times the text he produced the following summer, while taking a rest cure in the Alps, finally arriving at the resigned impression that he lacked the "talent of the storyteller" to "conjure a genuine picture of life," and abandoning the text he had produced in that first version.

The notes and stories about his life that Schrödinger wrote in those final months make for curious, at some points even bizarre, reading. Though as a young man Schrödinger dreamt of becoming a poet and even wrote his own poems;[4] though he read the canonical authors over and over, citing them with enthusiasm, what strikes the reader—the singular impression made by the text—is not exactly its literary quality. It is very difficult to grasp why Schrödinger works on us so queerly. The "autobiographical sketches"[5] may be so fascinating because, on the one hand, they assemble, with particular expressive intensity, a hodgepodge of elements rather than a comprehensive picture of an individual world or even a personal résumé, while on the other hand—despite the insistent absence of all things private—they convey an impression of Schrödinger's thought and feelings, which in their immediacy are almost uncomfortably affecting. That his personality was complex to the point of eccentricity[6] is beside the point (and is in any case rendered in living color by his autobiographical text).

Thus I would like to leave aside now the simple association between personality and text in order to hazard the claim—and then make it essentially plausible—that Schrödinger's recollections, written free of all preconception, convention, and commitment vis-à-vis the reader, make visible a kind of third dimension, an individual habitus of perception and experience; and that this habitus may constitute the matrix underlying his scientific discoveries and intuitions. Putting it another way, we could simply ask whether that very quality of personality which his contemporaries regarded as either uniquely charming or as absolutely insufferable, and which the reader of his autobiography finds so unsettling, may have been the unique structure that gave rise to Schrödinger's observations of natural phenomena and his "risky thinking."[7] Such intuitions certainly do not allow for any scientific or empirical proof. Yet if we can only be content with empirically proven answers, we will never be allowed to truly interrogate the conditions of the perhaps unique capacity for innovation that characterizes Erwin Schrödinger's work.

～

The two opening sentences of *My Life*, in different respects, already hold a surprise for the reader:

> I have lived the last years of my life separated from my closest, indeed my really very close friend. (Perhaps that is the reason why I have been reproached, increasingly often, with having no feeling for friendship, but merely for romance). (13)

What is the significance here of Schrödinger's shift from the superlative "my closest . . . friend" to "my really very close friend"? It seems what is likely happening in this move is the rejection of the established conventional formula ("Everyone has a closest friend") in favor of the individualistic reflection that his closest friend was in fact only "very close" to him. Above all, however, it is surprising that Schrödinger actually begins his memoir by reflecting on that friend; begins it, moreover, by expressing regret that their lives departed from their parallel course. It seems possible then that the reader would follow this friendship and the impossibility of its fulfillment as a central motif of Schrödinger's reflections, even as a trauma. But this is not the case. It is only to the reproach that he understood only romance, never friendship, that the narrator will return—and then only in the final passage of his text, and only indirectly.

What becomes apparent as one moves through Schrödinger's text—against all conventions of readerly expectation—is the nullification of any kind of narrative linearity and the complete absence of any hypotactical significance. Schrödinger's sketches consist of ten relatively vague thematic units[8] arranged according to no discoverable principle of organization and thematizing the relationship with his friend; his time as a student at the University of Vienna; memories of the last years of his parents' lives, immediately following the Great War; his childhood and youth, specifically the education provided by his parents; summer holidays during his childhood; further recollections of his studies; early experiences of the Great War; and later experiences of that war, stretching to his marriage and his appointment at Jena in 1920—at which point, surprisingly, Schrödinger formulates a "chronologically-organized biography" divided into six "periods," and concludes with a brief look back, which he composed in November 1960. Attention to the transitions between these sections soon reveals how often a motif or key word is the obvious catalyst for these haphazard changes of topic.

In this way, Schrödinger moves from the discussion of his lingering feelings of guilt about his parents to a consideration of their greatest credit, that is, the education they gave him; likewise, he finds in the formal features of his education the link to the musings that follow on the happy educational travels of his childhood summers. At times, however, the change is wholly abrupt; thus, after the first section, which concludes with memories of his friend's relations, Schrödinger jumps to his years as a student.

The sole narrative principle seems to be a refusal to submit the agility of his shifts in attention to external restrictions. This is made clear by the narrator's surprising introduction to the conventionally organized portion of his work: "A chronologically-organized biography—whether autobiographical or not—is, at least to my mind, one of the most boring things, since in nearly everyone's life there are at most a few noteworthy experiences or observations, and only very seldom does the entire succession of events in his history seem important to anyone other than the individual himself" (p. 35).

This freedom to turn at any time to the emerging "experiences and observations" that struck him as particularly interesting, was made possible through the complete eradication of narrative linearity. Schrödinger was never led astray by guidelines or conventional parameters, either personally or professionally. Nothing could dissuade him from savoring his always prolonged holidays to the fullest; he adored the Vienna Theater, above all Grillparzer's dramas, though he felt no Viennese duty to develop a liking for classical music.[9]

Admittedly, the situations and perceptions so surprisingly invoked by Schrödinger exist for him in a concreteness of strong contours, which often work on the reader in ways that seem excessive and therefore dysfunctional. What significance can anyone other than Schrödinger himself attribute to the fact that he rejoined his "really very good" friend—who in the text is called by the name "Fränzel"—in 1956 in Vienna at Pasteurgasse 4 (14)? Or that an "ideological chasm" loomed between Schrödinger and Fränzel's brother, though that same brother had been his guide "as a young man" on an expedition to a mountain called

"Einser" in the Sexten Dolomites (15)? The text abounds with
names of people and places of even less significance, which is in
itself revealing, since names—as opposed to nouns—always refer
to individual objects in the world. This is Schrödinger's obses-
sion: the things of the world or at any rate his first encounter
with them; to perceive or to internalize them individually, and
then to represent them as individual phenomena. He was un-
easy with the generalizing and typifying gaze. Thus (despecified)
concepts never seemed to be able to mediate between objects
in their concreteness and Schrödinger's experience. He often referred to people and things from his own world that had not yet
been mentioned in the text as if he could presuppose a familiarity of the reader with the object it references: "I heard Jerusalem
speak about Spinoza—whoever has heard him will not forget it.
He spoke about much, about Epicurus *ho thanatos oudèn pros hê-
mas*, about that wonder [*thaumazein*] that was the starting point
of every philosopher and others" (30f). The reader senses how
detailed and contoured the character of Schrödinger's memories
of the philosopher Jerusalem must have been, though he cannot
follow the image even one step further. Such idiosyncrasies of
precise perception and recollection point to how the reader's
comprehension is interrupted, as when Schrödinger—again
without preparation or commentary—uses a noun ("Phoenix-
Jew") to designate the tenant of his parents' apartment, a term
whose appearance within the Viennese dialect was quite obviously anchored to, and conditioned by, a distinct historical moment: his mother having to "give way to a financially sound
renter, so that she took care of my future father-in-law, in the
form of a rich Phoenix-Jew, in the most amiable way" (22).

Sometimes the same idiosyncratic language creates an almost
childish and yet more pleasurable effect. Before Schrödinger and
his wife fled Nazi Germany in 1933, "from Malcesine via Ber-

gamo, Lecce, St. Gotthard, Zürich, Paris to a Consul Solvay in Brussels," and from there finally to Oxford, they had purchased "a small BMW" for their escape (36).[10] In the text, this car is constantly referred to as the "brave grayling" (*braver Grauling*), so the two words quickly give the impression of a proper name, which has an anthropomorphizing effect. The small BMW again shows up several times in letters from a trip through Spain the couple took during the same year.[11] And in 1938, when Schrödinger and his wife had to flee the Nazis once more (this time from Graz), having to leave behind the "brave Grayling that would have been too slow," the reader feels the loss almost as keenly as he would the death of a beloved character in a children's book.

And yet in a few stylistically very successful passages of his text, Schrödinger does not merely refer to or name the objects in the act of remembrance but opens up his narrative to more incisive description. This seems to be the case particularly when the recollection is related to a very early phase of his life and is more difficult to retrieve. In these moments, objects and experiences come to the fore and coalesce into some kind of 'protophysics', as in a passage evoking a summer visit to his mother's family in England:

> Regular summer travels contributed not only to enriching my life, but also facilitated the progress of my thinking. I remember a trip to England, to my mother's English relatives, one year before I entered middle school. On the wide beach of Ramsgate, I liked to ride the donkey, and I learned to ride my bike there, too. I was deeply impressed with the tides; the little huts for changing clothes were put on wheels, and a man with a horse was constantly pulling them up or down the beach, depending on the movement of the tides. On the English channel I realized, for the first time, that one could see only the trails of smoke from far-away boats, whereas they themselves were hidden by the vaulting of the expanse of water. (27)

With a similar precision, Schrödinger writes about the light
in his parents' apartment:

> The spacious apartment [it actually was a double apartment]
> on the fifth floor of what was at that time a valuable apartment
> building in the city center, belonging to my mother's father, did
> not have electric light. Partly because my grandfather did not
> want to pay for the connection, partly because all of us, and my
> father in particular, had become used to the exquisite Auer- and
> Gräzin-light that we had been using already at a time when elec-
> tric lightbulbs were still expensive, reddish yellow night-lights, so
> no-one really wished to have them. (19)

That Erwin Schrödinger and the protagonists of his autobiographical narration entertained a special relationship with
some of these objects (as in 1920, when electric lightbulbs did
not seem to be the eventual replacement of gaslight for the
Schrödinger family) does not affect the concision and stability
of his world. What he perceives, differentiates, and describes are
things and human beings that manifest themselves in their singularity, in the distance at first and independent of the reactions
of their beholders. To see such singularity and to secure it against
the typifying tendency of our linguistic concepts was the obses-
sion of Schrödinger's perception of the world.

But whenever he starts thinking about such things, people, or
even institutions, Schrödinger lets himself indulge in potentially endless turns of reflection and interpretation that, however,
never seem to relativize the contours of his references. Such a
move pervades his introductory sentence—for us an already familiar formulation—about his "really very close friend" Fränzel.
And it is this move that probably causes the strange feeling that
*My Life* leaves with the reader, right from the beginning. Just a
few sentences tell us about the intense, seemingly "philosophi-

cal" conversations through which the two grammar school pupils constituted their friendship. Schrödinger mentions several
encounters later in life that were invariably disappointing and
certainly did not fulfill the promise of their early meetings: "We
exchanged only rare, never long, and hardly exhaustive letters.
It seemed to me that he had changed a lot over the years" (13);
in 1956 he met "Fränzel, but only for fifteen minutes; so it does
not really count" (14). "Two years later he died unexpectedly,
without our having seen each other again" (13). As if it could
not be true that his and Fränzel's closeness had been no more
than a youthful episode, Schrödinger remembers—in very dry
language—a time when he visited two of his friend's brothers
in Krems and Klagenfurt later on, and how he was unable to revive the lost feeling of familiarity he was looking for. However,
despite this polyperspectival succession of contradictory experiences, the image and also the primacy of his "really very close
friend" Fränzel in *My Life* remains untouched.

In the same introductory passage, a succession of only two
sentences illustrates even more directly how Schrödinger's multiple perspectives in diverse interpretations displace each other in
fast progression. He recalls how Fränzel and he thought of their
own juvenile conversations as highly inventive, only because they
had not yet been exposed to the tradition of Western philosophy
at school. The institutional power of the Catholic faith and the
Church in Austria was too strong: "The reason was that philosophical questions were generally avoided in the public schools,
in deference to the religious education teachers who decide all of
these things with higher authority and do away with it, because
it is cumbersome for them" (13). That Schrödinger regarded this
situation and its consequences as the reason for his own alienation
from religion is not really surprising; it does surprise, however,
when, commenting all of a sudden from the opposite direction,

he admits that religion had never actually harmed him: "This is the main reason for my opposition to religion, which, however, has never really done me any harm."

If one can say that the "world" as a "levitating texture"[12] is the irreducible structure of objects and their multiple interpretation through different human beings, then it was Erwin Schrödinger's style of world experience and world constitution that stood out because of its particular dynamics; or as one would say today, because of its specific sustainability. As much as the objects keep their complex individual contours in this style of experience, the series of new perspectives and interpretations by which he allows the objects to appear before him will never come to an end.[13] Because of this constant, potential infinity of perspectives, Schrödinger naturally experiences and imagines the objects through interpretations, which are far from the standard interpretations established within the respective historical and cultural daily routine. Included here are interpretations that are expelled from the constructs of daily life because the risk of ambivalence and even grotesque misunderstanding is inherent to them. But it is precisely this risk that intellectual and scientific thought has to face as long as it is supposed to be innovative. Scientific thought is entitled to run this risk all the more, since its institutional distance from daily life (which is often reproachfully called science's Ivory Tower) also protects science from contaminating daily life with unacceptable risks.

It is characteristic of this polyperspectivism—for his particular version of risky thinking—that Schrödinger repeatedly refers to people and institutions as "unique," only to integrate them in a second step into a whole group of similar people and institutions that may be of the same standing, or coequal: "Intellectually, there is no one who influenced me more than Fritz Hasenöhrl [his most important academic teacher], except perhaps for my father Rudolf—over the years we lived together, through the sustained conversations on basically everything he was interested in, and that was a lot" (15f). Similarly, it is the University of Vienna that seems to him the center of specific fields of research, even if, at second glance, it becomes clear that this center had moved to other cities: "Atmospheric electricity and radioactivity, for example, which both had their beginnings in Vienna, were taken from us. And whoever wanted to really contribute to these fields had to follow them, like Lise Meitner, who had to move from Vienna to Berlin for that purpose" (17). On another occasion, he describes a situation of scientific consensus, emphasizing the aspect of collectivity in it, only to suddenly switch to the perspective of his father and his own youth, embodying this consensus in only one scientist: "On one thing we all agreed—by *all* I am thinking particularly of one of my father's scientist friends that I knew and liked the best, the court counselor Anton Handlirsch [zoology, fossil insects] of the courtly natural history museum—I want to say, we agreed that the theory of development could be justified only causally, not teleologically" (26).

In such open sequences of interpretation, Schrödinger seems to be at least one step—and one perspective—beyond the standard opinions that determine erudite and specialized discussions. After having confirmed, in the passage about his own education, the common opinion that both genes and the "cultural and human environment" are crucial for educational success, he complements this necessary convergence with a third aspect, "home life": "That chromosomes plus school are not everything, but that there is also home life, to prepare the soil for the seed that the school is supposed to sow, that is something that those people, who are determined to give access to higher education only to children with a lower class background, ignore" (25).

Usually Schrödinger starts his polyperspectival sequences with the interpretation and the aspects that one is most likely to expect, that is, the most normal ones. Then he suddenly opens our perception to more surprising, more eccentric possibilities of experience. And yet, at a particularly significant passage of his autobiographical sketch, he reverses this order, telling us first about a nightmare that haunted him throughout his adult life, which he understood as originating in a sense of guilt toward his parents because he felt he had not cared for them enough in their later years; then suddenly turning this perspective up-side down when he narrates, in an almost aloof fashion, that other than that incident he is not usually haunted by such feel-ings: "I understand this nightmare as the result of my bad con-science about my neglectful behavior toward my parents during the years 1920/21. Otherwise, I am mostly untroubled by such nightmares; things gnaw at my conscience only rarely" (23).

Detached and unsentimental as he was, an untragical relationship to the events and constraints that might have affected or even imperiled his existence followed Erwin Schrödinger's habitus of polyperspectivism in his experience of the world. This distance gave him the freedom to engage with objects and people with peculiar intensity. He writes about the cause of death of his be-loved parents in laconic fashion—in brackets noting, in clinical-diagnostic terminology, "rapid decline through arteriosclerosis" (21) and "metastasis of the spine after a radical mastectomy for breast cancer in 1917" (23). Schrödinger records without com-ment, protest, or even an expression of sympathy the suicide of a valued colleague from Jena, who killed himself after the Nazi takeover: "My own position was helped along by a very nice couple, the Auerbachs (Jews), who helped us with the same heartfelt friendship as my boss Max Wien and his wife (anti-Semites, but more because of their background, not virulently). About the Auerbachs, we were informed after the takeover (1933) that they escaped the anticipated hardship and disrespect through suicide" (34).

Without a gesture of mediation or explanation, Schrödinger puts these irreconcilables—his sympathy for both his Jewish and his anti-Semitic colleagues—next to each other; he allows himself such paradoxical positions also in his daily life. With regard to World War I, he qualifies Austria's late and unsuccessful attempts to negotiate a separate peace treaty at first as "perfidious," and then regrets, in a truly paradoxical volte-face, the failure of these attempts: "The sojourn of the brother of our Empress Zita, Prinz Sixtus of Bourbon, had the goal of a separate peace treaty between the Austrian-Hungarian Empire and the Entente Cordiale, thus a malicious treason of Germany, which, however, unfortunately fell through" (33).

Schrödinger's unsentimental distance, coming from his endlessly varied interpretations, made him flexible in dealing with events in his personal and historical environment, events which most of his contemporaries would have experienced (and did) as irreversibly negative blows of fate. That his first academic job was as assistant to the physicist Franz Exner, and not to Fritz Hasenöhrl, whom he admired so much, left him retrospectively "deeply grateful," because it was this coincidence that forced him to become acquainted with measurement devices, techniques of measuring, and experimental work in a physics laboratory in general, none of which was his intellectual inclination: "Thus, today, I belong to those theoreticians who know from their own experience what it actually means to measure" (17). His reaction to the political changes in 1933 was one of composure, which then turned into resolution.[14] Although he had never shied away from speaking with a certain irony about Hitler, even in public, he did so without insistence; and he did not belong to the up-settingly small number of scientists who used their international reputations to support the protests during the first months of the Nazi regime, hopeless though they were. He simply left Germany, having asked for a temporary leave of absence from his academic duties, then enjoyed the summer holidays as he usually did (in 1933, in the company of many emigrant friends in South Tyrol). In December of that very year he accepted the Nobel Prize, perhaps his grand personal triumph. Five years later, after his second emigration (this time from Graz, where he had assumed a professorship in 1936), he found a new home in Dublin, where he would stay for more than seventeen years:

> The period from 1939 to 1956 I call the *long exile*, without, however, wanting to emphasize the bad taste of these words, because I had a very, very good time there. Without it, I would have never gotten to know this beautiful island; nowhere else could we have lived through this horrible war so uncomplainingly that I almost feel ashamed of it. [. . .] We often tell ourselves "we owe this to the Führer." (38)

It was in February 1943, in Dublin, that Schrödinger held the three Trinity lectures entitled "What Is Life?"[15] which appeared shortly after as a book, and would become an important stimulus for the upcoming field of biogenetics.

∼

Rarely does Schrödinger disrupt the sustained sequence of perspectives and interpretations of his autobiographical narration. In passages where he does disrupt that narration, one cannot help but get the impression that he (in an involuntarily almost comic way) is looking intensely for highly sophisticated, even over-the-top formulations meant to fix a definite idea within a gesture of conceptual eccentricity. One example would be the sentence that follows the proud conclusion that he was one of the few physicists in his field with a certain competence in measuring techniques: "*Mir deucht, es stünde besser, wären derer mehr*" [It seems to me all would be better if we had more scientists of my sort] (17). Similarly, the end of his reflection on components and principles of education appears forced: "If only teachers of all ranks—let alone parents—took the fact to heart that mutual goodwill will always be the necessary precondition for the success of the influence they exert on these young humans that are entrusted to them!" (29). Such sentences, meant to be quoted, and finishing with an exclamation point, were not Schrödinger's strength.

Erwin Schrödinger's strength, rather, was the description of the convergence of an individual perspective on objects, and the endless sequence of interpretations that often provoked counterintuitive experiences, allowing for a certain distance, composure, and flexibility regarding the world but also for a passionate participation in the world. It was the habitus of Schrödinger's essays and talks to alternate between 'realistic' passages, into which he crammed his idiosyncratic, world-constructing view; and cascades of consciously subjective speculation. It has been emphasized,[16] and I think rightly so, that this intellectual style reflects the influence of Schrödinger's most important Viennese teachers in the history of science, namely the principles of the realist and atomist Ludwig Boltzmann, and the phenomenological-epistemological reflections of Ernst Mach about world construction in experimental and theoretical physics. The colleagues and students of Boltzmann and Mach had to live with the competition between them; it was probably this competition that drove Boltzmann to commit suicide, in 1906. Schrödinger's admiration for both of his teachers was so great that he did not find their worldviews incompatible.[17] Without Mach, Schrödinger prob-

ably would not have kept his lifelong enthusiasm for Schopenhauer and for Hindu philosophy.[18] Without Boltzmann's example of how to leave phenomena as they are and return to them,
amazed, again and again, Schrödinger would have easily been
lost within his own multifaceted speculations. He described his
own work in the emerging field of biochemistry as the "half-
blind, intuitive groping that, however, leads ahead" (31). Such
groping is not possible without faith in actual objects.

In the second sentence of his autobiography, Schrödinger
mentions rather casually the reproach leveled against him that
"he would have no sense for friendship but just for flirtatious-
ness" (13). Only a few lines before the end of the text, he finally
picks up on this remark:

> I am not a gifted enough narrator to create a true life image—
> but neither do I have the possibility, since I had to leave out my
> relationships to women, which on the one hand caused a serious
> gap, but on the other hand seemed to be necessary, first because
> of the gossip they would have generated; secondly, since they
> would not have been interesting enough; and thirdly, because
> nobody can or is allowed to really be sincere and true in such
> things. (40)

These are not the self-glorifying memories of an old man; Erwin Schrödinger seems, on the contrary, to play down the reality
of his past in an almost grotesque manner. Since his adolescence,
during a marriage to Annemarie Bertel, which in its own way
was "happy," he had enjoyed every erotic adventure that life offered; these inspired him. Inevitably, we associate this uncondi-
tional polygamous lifestyle with the multiple perspectives of his
intellectual style; all the more so since Schrödinger, in his daily
life, was flexible enough to find livable solutions for the (according to a bourgeois understanding) "most impossible" con-
stellations. He encouraged his childless wife Annemarie to have

her own love affairs; and it was Annemarie herself who finally helped him to rear his daughter Ruth, born in 1934 to Hilde March, the wife of his colleague Arthur March. His daughter with Sheila May Greene, his Irish lover (whom he liked to remember in connection with the Trinity lectures[19]), grew up as the child of a man from whom Sheila got divorced soon after. And one year after her birth, in June 1946, Schrödinger had another daughter, with Kate Nolan.

Schrödinger got the decisive idea for wave mechanics (which eventually brought him the Nobel Prize) in the last days of 1925, during a skiing holiday in Arosa, where a for-once yet-unknown lover[20] accompanied the thirty-eight-year-old. It was speculated that each of his important discoveries was bound to a short but passionate love affair. Perhaps this thought is a little too simplistic—but he certainly needed (in the same way he always returned from his many open speculations to the concrete objects in which they had originated) individual and, for him, unique people, in which the world differentiated itself in a way that allowed him to react to its objects in their multiplicity. During the last hours before Erwin Schrödinger's death, on January 3, 1961, it was his wife, Annemarie, who held his hand.[21]

*Translated by Lisa Ann Villareal and Fabian Goppelsröder*

# Notes and References

## Notes to Gumbrecht: Introduction

1. Quoted in: Walter Moore, *A Life of Erwin Schrödinger* (Cambridge University Press, 1994), 101, 106, 169.

2. I refer to that historical moment which Michel Foucault in *Les Mots et les choses* (Éditions Gallimard, 1966) called the "crisis of representation." For the basis of a reinterpretation of the "crisis of representation" as a collection of epistemological consequences that arose due to the emergence of the second-order observer as an institutional form, cf. the second chapter of my book *The Production of Presence* (Stanford University Press, 2004).

## Notes to Harrison: Schrödinger on Mind and Matter

1. For Bacon's remarks about wonder as "broken knowledge" and "contemplation broken off," see "The Advancement of Learning," in *The Works of Francis Bacon*, ed. J. Spedding, vol. 6 (Taggard and Thompson, 1863), 96; and op. cit., "Valerius Terminus of the Interpretation of Nature," 29. For a discussion of Bacon's theories about wonder, as well as an inspired treatment of the role of wonder in human knowledge, see Andrea Nightingale's essay "Broken Knowledge," in *The Re-*

*enchantment of the World*, ed. Joshua Landy (Stanford University Press, 2009).

2. *What Is Life?* (Cambridge University Press, 1967; 11th printing, 2004). This book brings together a series of public lectures Schrödinger originally delivered at Trinity College in 1943.

3. So as not to leave these terms suspended in enigma, here is how Schrödinger defines an aperiodic crystal:

> A small molecule might be called "the germ of a solid." Starting from such a small solid germ, there seem to be two different ways of building up larger and larger associations. One is the compara- tively dull way of repeating the same structure in three directions again and again. That is the way followed in a growing crystal. Once the periodicity is established, there is no definite limit to the size of the aggregate. The other way is that of building up a more and more extended aggregate without the dull device of repetition. That is the case of the more and more complicated or- ganic molecule in which every atom, and every group of atoms, plays an individual role, not entirely equivalent to that of many others (as is the case in a periodic structure). We might quite properly call that an aperiodic crystal or solid and express our hypothesis by saying: We believe a gene—or perhaps the whole chromosome fibre—to be an aperiodic solid. (60–61)

In the opening pages of *What Is Life?* Schrödinger practically declares that one need be a physicist to appreciate how exceptional is the very phenomenon of an aperiodic crystal, which is not dominated by the law of statistical averages in the way that inanimate solid matter is:

> The most essential part of a living cell—the chromosome fibre— may suitably be called an *aperiodic crystal*. In physics we have dealt hitherto only with *periodic crystals*. To a humble physicist's mind, these are very interesting and complicated objects; they consti- tute one of the most fascinating and complex material structures by which inanimate matter puzzles his wits. Yet, compared with the aperiodic crystal, they are rather plain and dull. The differ- ence in structure is of the same kind as that between an ordinary wallpaper in which the same pattern is repeated again and again

in regular periodicity and a masterpiece of embroidery, say a Raphael tapestry, which shows no dull repetition, but an elaborate, coherent, meaningful design traced by the great master. (5)

And here is how Schrödinger speaks of "negative entropy":

[A] living organism continually increases its entropy—or, as you may say, produced positive entropy—and thus tends to approach the dangerous state of maximum entropy, which is death. It can only keep aloof from it, i.e., alive, by continually drawing from its environment negative entropy—which is something very positive. . . . What an organism feeds upon is [not matter or energy but] negative entropy. Or, to put it less paradoxically, the essential thing in metabolism is that the organism succeeds in freeing itself from all the entropy it cannot help producing while alive. (71)

4. In his "Note to the Epilogue," Schrödinger acknowledges his debt to Aldous Huxley's *The Perennial Philosophy*, which he calls "a beautiful book . . . singularly fit to explain not only the state of affairs [of human consciousness], but also why it is so difficult to grasp and so liable to meet with opposition" (90).

5. *Mind and Matter* accompanies *What Is Life?* in the edition cited above. Page numbers refer to that edition.

6. Edward O. Wilson, *On Human Nature* (Harvard University Press, 1978), 7.

7. *New York Times*, Op-Ed., Nov. 24, 2007.

8. In *Being and Time* Heidegger writes: "Before Newton's laws were discovered, they were not 'true'. . . . Through Newton the laws became true; and with them, entities became accessible in themselves to Dasein. Once entities have been uncovered, they show themselves precisely as entities which beforehand already were. Such uncovering is the kind of Being which belongs to 'truth'" (226–27).

## References for Laughlin: Schrödinger's Trouble

1. E. Schrödinger, *The Interpretation of Quantum Mechanics: Dublin Seminars (1949–1955) and Other Unpublished Essays* (Ox Bow Press, 1995).

2. E. Schrödinger, "Quantizierung als Eigenwertproblem," *Annalen der Physik* 79 (1926): 361; 79 (1926): 489; 80 (1926): 437; 81 (1926): 109.

3. D. R. Hartree, "The Wave Mechanics of an Atom with a Non-Coulomb Central Field," *Proceedings of the Cambridge Philosophical Society* 24 (1927): 89, 111.

4. E. A. Hylleraas, "Neue Berechnungen der Energie des Heliums im Grundzustande, sowie des tiefsten Terms von der Orth-Helium," *Zeitschrift für Physik* 54 (1929): 347.

5. J. C. Slater, "The Theory of Complex Spectra," *Physics Review* 34 (1929): 1293.

6. V. A. Fock, "Näherungsmethode zur Lösung des quantenmechanischen Mehrkörperproblems," *Zeitschrift für Physik* 61 (1930): 126.

7. H. A. Bethe and E. E. Salpeter, *Quantum Mechanics of One- and Two-Electron Atoms* (Academic, 1957).

8. W. Heitler and F. London, "Wechselwirkung neutraler Atome und homöpolare Bindung nach der Quantenmechanik," *Zeitschrift für Physik* 44 (1927): 455.

9. M. Born and R. Oppenheimer, "Zur Quantentheorie der Molekeln," *Annalen der Physik* 20 (1927): 457.

10. E. Hückel, "Quantentheoretische Beiträge zum Benzolproblem," *Zeitschrift für Physik* 70 (1931): 204.

11. P. A. M. Dirac, "On Quantization of Perfect Monatomic Gases," *Proceedings of the Royal Society of London* A 112 (1926): 661.

12. A. Sommerfeld, "Zur Elektronen Theorie der Metalle auf Grund der Fermischen Statistik," *Zeitschrift für Physik* 47 (1928): 43.

13. F. Bloch, "Über die Quantenmechanik der Elektronen in Kristalgittern," *Zeitschrift für Physik* 52 (1928): 555.

14. E. P. Wigner, "On the Interaction of Electrons in Metals," *Physics Review* 46 (1934): 1002.

15. N. F. Mott and H. Jones, *The Theory of the Properties of Metals and Alloys* (Oxford University Press, 1936).

16. N. F. Mott, "The Basis of the Theory of Electron Metals, with Special Reference to the Transition Metals," *Proceedings of the Royal Society of London* A 62 (1949): 416.

17. R. P. Feynman, "Space-Time Approach to Quantum Electrodynamics," *Physics Review* 76 (1949): 769.

18. J. Schwinger, *Selected Papers on Quantum Electrodynamics* (Dover, 1958).

19. G. Gamow, "Zur Quantentheorie des Atomkernes," *Zeitschrift für Physik* 51 (1928): 205.

20. E. Fermi, "Versuch einer Theorie der $\beta$-Strahlung," *Zeitschrift für Physik* 88 (1934): 161.

21. H. Yukawa, "On the Interaction of Elementary Particles," *Proceedings of the Physical-Mathematical Society of Japan* 17 (1935): 48.

22. N. Bohr and J. A. Wheeler, "The Mechanism of Nuclear Fission," *Physics Review* 56 (1939): 426.

23. K. Popper, *Quantum Theory and the Schism of Physics* (Routledge, 1992), 125.

24. E. Schrödinger, "Über den Comptoneffekt," *Annalen der Physik* 82 (1927): 257.

25. E. Schrödinger, "Verhalten des Starkeffekts bei plötzlichen Feldänderungen," *Zeitschrift für Physik* 78 (1932): 309.

26. E. Schrödinger and M. Born, "The Absolute Field Constant in the New Field Theory," *Nature* 135 (1935): 342.

27. E. Schrödinger, "Probability Relations Between Separated Systems," *Proceedings of the Cambridge Philosophical Society* 31 (1935): 555.

28. E. Schrödinger, "Phenomenological Theory of Supra-Conductivity," *Nature* 137 (1936): 824.

29. E. Schrödinger, "Mean Free Path of Protons in the Universe," *Nature* 141 (1938): 410.

30. E. Schrödinger, "Nature of the Nebular Red Shift," *Nature* 144 (1939): 593.

31. E. Schrödinger, "The Proper Vibrations of the Expanding Universe," *Physica* 6 (1939): 899.

32. E. Schrödinger, "The Union of the Three Fundamental Fields (Gravitation, Meson and Electromagnetism)," *Proceedings of the Royal Irish Academy* 49A (1944): 275.

33. E. Schrödinger, "Probability Problems in Nuclear Chemistry," *Proceedings of the Royal Irish Academy* 51A (1945): 1.

34. E. Schrödinger, "A Combinatorial Problem in Counting Cosmic Rays," *Proceedings of the Royal Society* A 64 (1951): 1040.

35. E. Schrödinger, "Was ist ein Naturgesetz?" *Nature* 17 (1929): 9.

36. E. Schrödinger, "Indeterminism and Free Will," *Nature* 138 (1936): 13.

37. E. Schrödinger, "World Structure," *Nature* 140 (1937): 742.

38. E. Schrödinger, *Über Indeterminismus in der Physik—Ist die Naturwissenschaft milieubedingt?* (Barth, 1932).

39. E. Schrödinger, *Science and the Human Temperament* (Norton, 1935).

40. E. Schrödinger, *What Is Life?* (Cambridge University Press, 1944).

41. E. Schrödinger, *Science and Humanism: Physics in Our Time* (Cambridge University Press, 1951).

42. E. Schrödinger, *Gedichte* (Küpper, 1949).

43. J. Honner, *The Description of Nature: Niels Bohr and the Philosophy of Quantum Physics* (Oxford University Press, 1987).

44. F. G. Major, *The Quantum Beat: Principles and Applications of Atomic Clocks* (Springer, 2007).

45. M. Born, "Zur Quantenmechanik der Stoßvorgänge," *Zeitschrift für Physik* 37 (1926): 863.

46. E. N. Lorentz, *The Essence of Chaos* (University of Washington Press, 1996).

47. W. Heisenberg, "Über quantentheoretische Umdeutung kinematischer und mechanischer Beziehungen," *Zeitschrift für Physik* 33 (1925): 879.

48. E. Schrödinger, "Über das Verhältnis der Heisenberg Born Jordanischen Quantenmechanik zu der Meinen," *Annalen der Physik* 79 (1926): 361.

49. N. Bohr, *On the Quantum Theory of Line-Spectra* (Dover, 2005; 1918–22 editions).

50. J. Mehra and H. Rechenberg, *The Historical Development of Quantum Theory, Vol. 5* (Springer, 2000), 821.

51. "Je mehr ich über den physikalischen Teil der Schrödingerischen Theorie nachdenke, desto abscheulicher finde ich ihn. Man stelle sich das rotierende Elektron vor, dessen Ladung ist mit über den ganzen Raum verteilt, der Achse in einer 4. und 5. Dimensions. Was Schrödinger die seiner schreibt ich über Anschaulichkeit Theorie . . .

finde es Mist. Die große Leistung der Schrödingerische Theorie ist die Berechnung der Matrixelemente."

52. "Jetzt benützen die verdammte Göttinger meine schöne Wellenmechanik zur Ausrechnung ihrer Scheiß-Matrixelemente."

53. D. Wick and W. Faris, *The Infamous Boundary: Seven Decades of Heresy in Quantum Physics* (Springer, 1998), vii.

54. M. Arndt et al., "Wave-Particle Duality of C60," *Nature* 401 (1999): 680.

55. L. D. Landau and E. M. Lifschitz, *Quantum Mechanics* (Elsevier, 1981).

56. W. Heisenberg, "Mehrkörperproblem und Resonanz in der Quantenmechanik," *Zeitschrift für Physik* 38 (1926): 411; 41 (1926): 239.

57. S. Epstein, *Wage and Labor Guilds in Medieval Europe* (University of North Carolina Press, 1991).

58. W. Moore, *Schrödinger: Life and Thought* (Cambridge University Press, 1989).

59. E. Schrödinger, "Die gegenwärtige Situation in der Quantenmechanik," *Naturwissenschaften* 23 (1935): 807, 823 844. [Reprinted as "The Present Situation in Quantum Mechanics," *Proceedings of the American Philosophical Society* 124 (1980): 323; and in *Quantum Theory and Measurement*, ed. J. A. Wheeler and W. H. Zurek (Princeton University Press, 1983).]

60. F. Capra, *The Tao of Physics* (Bantam, 1977).

61. J. Gribbon, *In Search of Schrödinger's Cat: Quantum Physics and Reality* (Corgi Adult, 1985).

62. C. C. Gerry and P. L. Knight, "Quantum Superpositions and Schrödinger Cat States in Quantum Optics," *American Journal of Physics* 65 (1997): 964.

63. J. R. Friedman et al., "Detection of a Schrödinger's Cat State in an rf-SQUID," *Nature* 406 (2000): 43.

64. J. Bub, "Some Reflections on Quantum Logic and Schrödinger's Cat," *British Journal for the Philosophy of Science* 30 (1979): 27.

65. R. A. Wilson, *Schrödinger's Cat Trilogy* (Dell, 1988).

66. U. K. LeGuin, "Schrödinger's Cat," in *The Compass Rose* (Harper and Row, 1982).

67. A. L. Fetter and J. D. Walecka, *Quantum Theory of Many-Particle Systems* (Dover, 2003).

68. P. W. Anderson, "Plamons, Gauge Invariance and Mass," *Physics Review* 130 (1963): 439.

69. A. A. Abrikosov, L. P. Gor'kov, and I. Y. Dzyaloshinskii, *Methods of Quantum Field Theory in Statistical Mechanics* (Dover, 1975).

70. P. A. Lee and T. V. Ramakrishnan, "Disordered Electronic Systems," *Review of Modern Physics* 57 (1985): 287.

71. H. P. Stapp, "Light as a Foundation of Being," in *Quantum Implications*, ed. G. J. Hiley and F. D. Post (Taylor and Francis, 2007), 257.

72. *King Lear*, 3:2.

## Notes to Hendrickson:
## Exorcizing Schrödinger's Ghost

This manuscript grew out of the intensely stimulating cross-disciplinary environment of the Stanford Philosophical Reading Group, an atmosphere generated and sustained by the leadership of Sepp Gumbrecht and Robert Harrison. My thanks to Helga Wild and Niklas Damiris for their insights. I am indebted more than I can say to Charitini Douvaldzi for her incisive critical commentary and careful reading throughout all stages of the preparation of this manuscript.

1. The history of Schrödinger's and other physicists' contributions to the developing field of molecular genetics is well-worked territory (Keller 1990, 1995; Domondon 2006; Moore 1989; Olby 1971; Pauling 1987; Perutz 1987; Rosen, "The Schrödinger Question, What Is Life?" 2000; Symonds and Delbrück 1988; Yoxen 1979).

2. The historian Lily Kay provides a convincing deconstruction of what she terms the "Founding Father Narrative." In particular she argues that the attribution of information age concepts and language to Schrödinger is anachronistic (Kay 2000).

3. There are several excellent histories of classical and molecular genetics: Olby (1994); Judson (1996); Keller (1995, 2000); Carlson (2004); Morange (1998); Schwartz (2008).

4. At the beginning of the twenty-first century, it is very difficult, for reasons presented later in this essay, to provide a coherent defini-

tion of 'gene'. 'Gene', as we shall see, is used in at least two distinct senses. In one sense, it is the gene of Mendel and Morgan, that is, an abstract theoretical construct that points to an *observable trait* (say, flower color) which can assume different forms (purple flowers, white flowers). 'Gene' acquired a second sense in the decades after Schrödinger's book; in this sense it denoted a segment (continuous or discontinuous) of DNA that codes for a messenger RNA (more on this later). For this molecular version of 'gene' I will use $G_{Mol}$. Moss, who originally made these distinctions, uses Gene-P (P for preformationist) and Gene-D (D for developmental resource). I expand on his meaning in my concluding comments (Moss 2003, 2008).

5. There are different kinds of chemical bonds, which vary in strength (Watson et al. 2004). Covalent bonds are the strongest, while hydrogen bonds, electrostatic bonds, and van der Waals bonds are weaker. The double helix backbone is made up of covalent bonds; Watson-Crick base pairing is accomplished by weaker hydrogen bonding.

6. Schrödinger's knowledge of biochemistry was minimal. He was also unacquainted with the state of bioenergetics in the 1940s. The reactions to the bioenergetics discussion in *What Is Life?* ranged from calling it "appalling" (Perutz 1987; Pauling 1987) to seeing it as the work of genius presaging non-equilibrium thermodynamics (Schneider and Kay 1995; Schneider and Sagan 2005; Weber 2008).

7. Textual-semantic (or semiotic) and computer metaphors, novel in the 'Heroic Age', are now entrenched in the contemporary molecular biology discourse. I will use them, without quotation marks, in this spirit without intending to deny their metaphorical status or their potency.

8. The prokaryote-eukaryote distinction is conventional but outmoded in ways not relevant to this presentation (Woese 2004).

9. That said, it is clear today that bacteria are far from simple in many other respects (Zimmer 2008; Harold 2001; Shapiro 2007).

10. Reanalyzing bacteria in the postgenomic age has made it clear that, their simple anatomy notwithstanding, bacteria accomplish compartmentalization in other complex ways (Zimmer 2008; Harold 2001; Shapiro 2009).

11. *Transposons*, or 'jumping genes', are sequences of DNA that

can change their position in a cell's genome. Depending on where a transposon lands, it may lead to a mutation.

12. This term can lead to confusion. *Epigenesis* denotes an embryologic theory about the origin of form and is roughly synonymous with self-organization (Maienschein 2005).

*Epigenetic* also refers to stable modification of chromosomes that do not involve changes in the underlying DNA sequence. This may be accomplished by methylation of cytosine nucleotides and acetylation of histones (the chromosomal proteins ensheathing the DNA). Epigenetic control is essential to cell differentiation, the focused and selective expression of the genome in a particular context (say, making a breast duct lining cell).

13. The meaning of 'information' in biology is, currently, as problematic as the meaning of 'gene'. These issues are discussed in Godfrey-Smith (2007); Godfrey-Smith and Sterelny (2007); von Baeyer (2004); Jablonka and Lamb (2005); Jablonka (2002); Maynard Smith (2000).

14. A large literature on the conceptual status of 'gene' argues for either fashioning a more satisfactory definition or eliminating the term altogether (Griffiths and Stotz 2006, 2007; Keller 2000; Keller and Harel 2007).

15. A particularly spectacular example of this are the NCAM (for the neural cell adhesion molecule) proteins crucial in the development of the nervous system (Moss 2003, 2008).

16. For a critique of biotechnology see Wilkins's (2007) and Heng's (2008) comments. Pisano (2006) provides a critique from a business perspective.

17. A good place to start on medical genetics references is Scriver and Waters's (1999) discussion of phenylketonuria, and Weatherall's (2001) on thalassaemias. Weatherall is one of the world's experts on hemoglobin disorders.

18. Greaves (2000) provides a popular, up-to-date perspective on cancer and its history. Weinberg, who discovered one of the first tumor suppressor genes, has a recent, more technical account of cancer (2006). Morange (2002, 2007) and Moss (2003) review the history of theories of cancer causation and current cancer research. Fujimura (1996) explores the sociohistory of cancer research, while Heng (2007) and

Kitano (2004) provide a PSP critique of cancer genomics. The Cancer Genome Atlas (TCGA) results are detailed in several recent publications (Lin et al. 2007; Lin and Sjoblom 2007; Sjoblom et al. 2006; Wood et al. 2007).

19. The Second Law can be reformulated as "Nature abhors a gradient"; more than the Second Law being permissive of self-organization, self-organization actually promotes the fulfillment of the law's requirements (Schneider and Sagan 2005; Schneider and Kay 1995; Wickens 1987).

20. Now called Gibbs free energy, or usable energy.

21. Systems biology may be found treated in various publications (Fujimura 2005; O'Malley and Dupre 2005; Westerhoff and Palsson 2004; Morange 2006; Gilbert and Sarkar 2000; Greenspan 2001; Keller 2005; Strohman 1997; Boogerd et al. 2007; Kirschner 2005).

22. The fingerprints of a person's right and left hands differ despite the fact that the cells involved had the same genetic makeup and their development occurred in the same uterine environment. These differences ultimately are explained by the random, unequal distribution of the small number of biomolecules in dividing cells. Developmental noise denotes this contingent phenomenon (Lewontin 2004).

23. Thompson (2007); Rosen 1991; Rosen, *Essays on Life Itself* (2000); Mikulecky 2000.

24. Further discussion of the need for new laws to account for life include Kauffman (2000); Rosen, *Essays on Life Itself* (2000); and Érdi (2008).

25. Moss (2003) offers a detailed discussion of the oncogene theory and its twenty-first-century alternatives.

26. Surprisingly, the very expression 'Master Molecule', to this day the identifying banner of the genocentric perspective, was first used critically (if not ironically) by David Nanney. Equally surprising, in 1956 it was presented as one of the two possible ways of conceptualizing the organism. From its very beginning, it was explicitly cast in sociopolitical terms. Keller (2002) quotes Nanney:

> We may crudely locate the source of this tension in the difference between two leanings, one toward genetic control and the other toward cellular regulation. David Nanney's description of

these two conceptual bents, offered at a 1956 meeting on "The Chemical Basis of Heredity," has become something of a classic among historians of biology, and I quote it here: "The first of these we will designate as the 'Master Molecule' concept. This concept presupposes a special type of material, distinct from the rest of the protoplasm, which directs the activities of the cell and functions as a reservoir of information. In its simplest form the concept places the 'master molecules' in the chromosomes and attributes the characteristics of an organism to their specific construction; all other cellular constituents are considered relatively inconsequential except as obedient servants of the masters. This is in essence the Theory of the Gene, interpreted to suggest a totalitarian government. . . . The second concept . . . we will designate as the 'Steady State' concept. By the term 'Steady State' we envision a dynamic self-perpetuating organization of a variety of molecular species which owes its specific properties not to the characteristics of any one kind of molecule, but to the functional interrelationships of these molecular species. Such a concept contains the notion of checks and balances in a system of biochemical reactions. In contrast to the totalitarian government by 'master molecules', the 'steady state' government is a more democratic organization, composed of interacting cellular fractions operating in self-perpetuating patterns." (Keller 2002: 150ff)

27. The facetious designation 'molecular vitalism' has been suggested (Kirschner, Gerhart, and Mitchison 2000).

28. A large literature deals with the essential role of models and metaphors in science (Giere 2006; Frigg and Hartmann 2006; Bradie 1998, 1999).

29. See Kirschner, Gerhart, and Mitchison (2000) for a discussion of the dissimilarities between man-made machines and molecular machines (macromolecular assemblies).

30. See particularly works of Lewontin and colleagues (Lewontin, "Organism and Environment" 1982; Lewontin, Rose, and Kamin 1984; Lewontin 1993; Lewontin, *The Triple Helix: Gene, Organism and Environment* 2000; Bradie 1999; Fracchia and Lewontin 2005; Levins

and Lewontin 1985; Lewontin, *It Ain't Necessarily So: The Dream of the Human Genome and Other Illusions* 2000); and works of Denis Noble (Noble 2006; Noble, "Claude Bernard, the First Systems Biologist, and the Future of Physiology" 2008; Noble "Prologue: Mind over Molecule: Activating Biological Demons" 2008).

31. There is an extensive literature about the public rhetoric surrounding molecular genetics (Nelkin and Lindee 2004; Hubbard and Wald 1999; van der Weele 2004; Commoner 2002). An instructive metaphor for this complex and changing blend of public discourse is a "Theater of Genetics," a theater of representation. Van Dijck (1998) has structured an entire book about 'public gene talk' employing this trope.

## References for Hendrickson:
## Exorcizing Schrödinger's Ghost

Atlan, H., and C. Bousquet. *Questions de vie*. Paris: Seuil, 1994

Bailey, J. E. "Lessons from Metabolic Engineering for Functional Genomics and Drug Discovery." *Nature Biotechnology* 17 (1999): 616–18.

Boogerd, F. C., et al. *Systems Biology. Philosophical Foundations*. Amsterdam: Elsevier, 2007.

Bradie, M. "Explanation as Metaphorical Redescription." *Metaphor and Symbol* 13 2 (1998): 125–39.

———. "Science and Metaphor." *Biology and Philosophy* (1999): 159–66.

Cairns, J. *Cancer, Science and Society*. San Francisco: Freeman, 1978.

Carlson, E. A. *Mendel's Legacy. The Origin of Classical Genetics*. Cold Spring Harbor, NY: Cold Spring Harbor Laboratory Press, 2004.

Caruso, Denise. "Re:Framing; a Challenge to Gene Theory, a Tougher Look at Biotech." *New York Times*, July 1, 2007.

Cohen, I. R., and H. Atlan. "Genetics as Explanation: Limits to the Human Genome Project." *Encyclopedia of Life Sciences*. London: Macmillan, Nature Publishing Group, 2005.

Commoner, B. "Unraveling the DNA Myth. The Spurious Foundation of Genetic Engineering." *Harper's* Feb. (2002): 39–47.

Cornish-Bowden, A., et al. "Beyond Reductionism: Metabolic Circularity as a Guiding Vision for a Real Biology of Systems." *Proteomics* 7 6 (2007): 839–45.

Delbrück, M. "Aristotle-totle-totle." *Of Microbes and Life*. Eds. Monod, J., and E. Borek. New York: Columbia University Press, 1971. 54–55.

Domondon, A. T. "Bringing Physics to Bear on the Phenomenon of Life: The Divergent Positions of Bohr, Delbrück, and Schrödinger." *Studies in History and Philosophy of Biological and Biomedical Sciences* 37 (2006): 433–58.

Érdi, P. *Complexity Explained*. Berlin: Springer, 2008.

Falk, R. "What Is a Gene?" *Studies in History and Philosophy of Science* 17 2 (1986): 133–73.

Fracchia, J., and R. Lewontin. "The Price of Metaphor." *History and Theory* 44 (2005): 14–29.

Frigg, Roman, and Stephan Hartmann. "Models in Science." *Stanford Encyclopedia of Philosophy*. Stanford, CA: Center for the Study of Language and Information, Stanford University, 2006. <http://plato.stanford.edu/entries/information-biological/>

Fujimura, J. *Crafting Science: A Sociohistory of the Quest for the Genetics of Cancer*. Cambridge, MA: Harvard University Press, 1996.

———. "Postgenomic Futures: Translations Across the Machine-Nature Border in Systems Biology." *New Genetics and Society* 24 2 (2005): 195–225.

Gerhart, J., and M. Kirschner. "The Theory of Facilitated Variation." *Proceedings of the National Academy of Sciences USA* 104 Suppl 1 (2007): 8582–89.

Giere, R. N. *Scientific Perspectivism*. Chicago: University of Chicago Press, 2006.

Gilbert, S. F., and S. Sarkar. "Embracing Complexity: Organicism for the 21st Century." *Developmental Dynamics* 219 1 (2000): 1–9.

Godfrey-Smith, P. "Information in Biology." *The Cambridge Companion to the Philosophy of Biology*. Eds. Hull, D. L., and M. Ruse. Cambridge, UK: Cambridge Companions to Philosophy, Cambridge University Press, 2007. 103–19.

Godfrey-Smith, P., and K. Sterelny. "Biological Information." *Stanford*

*Encyclopedia of Philosophy.* Stanford, CA: Center for the Study of Language and Information, Stanford University, 2007. <http://plato.stanford.edu/entries/information-biological/>

Goldenfeld, N., and C. Woese. "Biology's Next Revolution." *Nature* 445 7126 (2007): 369.

Greaves, Mel. *Cancer: The Evolutionary Legacy.* Oxford, UK: Oxford University Press, 2000.

Greenspan, R. J. "The Flexible Genome." *Nature Reviews Genetics* 2 5 (2001): 383–87.

Griffiths, A. J. F. "Developmental Systems Theory." *Encyclopedia of Life Sciences.* London: Macmillan, Nature Publishing Group, 2002.

Griffiths, P. E., and K. Stotz. "Genes in the Postgenomic Era." *Theoretical Medicine and Bioethics* 27 6 (2006): 499–521.

Griffiths, P. E., and K. C. Stotz. "Gene." *The Cambridge Companion to the Philosophy of Biology.* Eds. Hull, D. L., and M. Ruse. Cambridge, UK: Cambridge Companions to Philosophy, Cambridge University Press, 2007. 85–102.

Haldane, J. B. S. "On Being the Right Size" (1926). In J. B. S. Haldane and J. M. Smith, *On Being the Right Size and Other Essays.* New York: Oxford University Press, 1985.

Harold, F. M. *The Way of the Cell: Molecules, Organisms and the Order of Life.* Oxford, UK: Oxford University Press, 2001.

Heng, H. H. "Cancer Genome Sequencing: The Challenges Ahead." *Bioessays* 29 (2007): 783–94.

———. "The Gene-Centric Concept: A New Liability?" *Bioessays* 30 2 (2008): 196–97.

Hubbard, R., and E. Wald. *Exploding the Gene Myth: How Genetic Information Is Produced and Manipulated by Scientists, Physicians, Employers, Insurance Companies, Educators, and Law Enforcers* (3rd ed.). Boston: Beacon Press, 1999.

Jablonka, E. "Information: Its Interpretation, Its Inheritance and Its Sharing." *Philosophy of Science* (2002): 578–605.

Jablonka, E., and M. Lamb. *Evolution in Four Dimensions: Genetic, Epigenetic, Behavioral, and Symbolic Variation in the History of Life.* Boston: Life and Mind: Philosophical Issues in Biology and Psychology, MIT Press, 2005.

Jacob, F. *La Logique du vivant.* Paris: Gallimard, 1970.

Jacob, F., and J. Monod. "Genetic Regulatory Mechanisms in the Synthesis of Proteins." *Journal of Molecular Biology* 3 (1961): 318–56.

Judson, H. F. *The Eighth Day of Creation. The Makers of the Revolution in Biology. Expanded Edition.* Cold Spring Harbor, NY: Cold Spring Harbor Laboratory Press, 1996.

Kauffman, S. *Investigations.* New York: Oxford University Press, 2000.

Kay, L. E. *Who Wrote the Book of Life: A History of the Genetic Code.* Stanford, CA: Stanford University Press, 2000.

Keller, E. F. "The Century Beyond the Gene." *Journal of Biosciences* 30 1 (2005): 3–10.

———. "The Disappearance of Function from 'Self-Organizing Systems'." *Systems Biology: Philosophical Foundations.* Eds. Boogerd, F. C., et al. Amsterdam: Elsevier, 2007.

———. "Physics and the Emergence of Molecular Biology: A History of Cognitive and Political Synergy." *Journal of the History of Biology* 23 3 (1990): 389–409.

Keller, Evelyn Fox. *The Century of the Gene.* Cambridge, MA: Harvard University Press, 2000.

———. *Making Sense of Life.* Cambridge, MA: Harvard University Press, 2002.

———. *Refiguring Life: Metaphors of Twentieth-Century Biology.* New York: Columbia University Press, 1995.

Keller, E. F., and D. Harel. "Beyond the Gene." *PLoS ONE* 2 11 (2007): e1231.

Kevles, D. J., and L. Hood. *The Code of Codes.* Cambridge, MA: Harvard University Press, 1992.

Kirschner, M. W. "The Meaning of Systems Biology." *Cell* 121 4 (2005): 503–4.

Kirschner, M., J. Gerhart, and T. Mitchison. "Molecular 'Vitalism'." *Cell* 100 1 (2000): 79–88.

Kirschner, M., and J. Gerhart. *The Plausibility of Life: Resolving Darwin's Dilemma.* New Haven, CT: Yale University Press, 2006.

Kitano, H. "Cancer as a Robust System: Implications for Anticancer Therapy." *Nature Reviews Cancer* 4 3 (2004): 227–35.

Levins, R., and R. C. Lewontin. *The Dialectical Biologist*. Cambridge, MA: Harvard University Press, 1985.

Lewontin, R. *Biology as Ideology*. New York: Harper Perennial, 1993.

———. "The Genotype/Phenotype Distinction." *Stanford Encyclopedia of Philosophy*. Stanford, CA: Center for the Study of Language and Information, Stanford University, 2004. <http://plato.stanford.edu/entries/genotype-phenotype/>

———. *Human Variation*. New York: Scientific American Library, 1982.

———. "In the Beginning Was the Word. Review of 'Who Wrote the Book of Life? A History of the Genetic Code' by Lily E. Kay." *Science* 291 5507 (2001): 1263–64.

———. "Organism and Environment." *Learning, Development, and Culture*. Ed. Plotkin, H. C. Chichester, UK: John Wiley, 1982. 151–70.

Lewontin, R. C. *It Ain't Necessarily So: The Dream of the Human Genome and Other Illusions*. Ed. Lewontin, R. C. New York: New York Review of Books, 2000.

———. *The Triple Helix: Gene, Organism and Environment*. Cambridge, MA: Harvard University Press, 2000.

Lewontin, R., and R. Levins. *Biology Under the Influence*. New York: Monthly Review Press, 2007.

Lewontin, R. C., S. Rose, and L. J. Kamin. *Not in Our Genes*. New York: Oxford University Press, 1984.

Lin, J., et al. "A Multidimensional Analysis of Genes Mutated in Breast and Colorectal Cancers." *Genome Research* 17 9 (2007): 1304–18.

Lin, J., and T. Sjoblom. "Genome-Wide Mutational Analyses of Breast and Colorectal Cancers." *Discovery Medicine* 7 37 (2007): 13–19.

Maienschein, J. "Epigenesis and Preformationism." *Stanford Encyclopedia of Philosophy*. Stanford, CA: Center for the Study of Language and Information, Stanford University, 2005. <http://plato.stanford.edu/entries/epigenesis/>

Maturana, H. R., and F. Varela. *Tree of Knowledge*. Boston: Shambhala, 1987.

———. *Autopoiesis and Cognition: The Realization of the Living*. Boston: Springer, 1991.

Maynard Smith, J. "The Concept of Information in Biology." *Philosophy of Science* 67 (2000): 177–94.

McClintock, B. "The Significance of Responses of the Genome to Challenge." *Science* 226 4676 (1984): 792–801.

Merlo, L. M., et al. "Cancer as an Evolutionary and Ecological Process." *Nature Reviews Cancer* 6 12 (2006): 924–35.

Mikulecky, D. C. "Robert Rosen: The Well-Posed Question and Its Answer—Why Are Organisms Different from Machines?" *Systems Research and Behavioral Science* 17 (2000): 419–32.

Moore, W. *Schrödinger, Life and Thought*. Cambridge, UK: Cambridge University Press, 1989.

Morange, M. "Cancer Research." *Encyclopedia of Life Sciences*. Chichester, UK: John Wiley, 2002. <www.els.net>

———. "The Field of Cancer Research: An Indicator of Present Transformations in Biology." *Oncogene* 26 55 (2007): 7607–10.

———. *A History of Molecular Biology*. Cambridge, MA: Harvard University Press, 1998.

———. "Post-Genomics, Between Reduction and Emergence." *Synthese* 151 (2006): 355–60.

Moss, L. "The Meanings of the Gene and the Future of the Phenotype." *Genomics, Society and Policy* 4 1 (2008): 38–57.

———. *What Genes Can't Do*. Cambridge, MA: A Bradford Book, MIT Press, 2003.

Murphy, M. P., and L. A. J. O'Neill, eds. *What Is Life? The Next Fifty Years: Speculation on the Future of Biology*. Cambridge, UK: Cambridge University Press, 1995.

Nelkin, D., and M. S. Lindee. *The DNA Mystique: The Gene as a Cultural Icon*. Conversations in Medicine and Society. Ann Arbor: University of Michigan Press, 2004.

Nicolis, G., and I. Prigogine. *Exploring Complexity. An Introduction*. New York: W. H. Freeman, 1989.

Noble, D. "Claude Bernard, the First Systems Biologist, and the Future of Physiology." *Experimental Physiology* 93 1 (2008): 16–26.

———. *The Music of Life: Biology Beyond the Genome*. Oxford, UK: Oxford University Press, 2006.

———. "Prologue: Mind over Molecule: Activating Biological Demons." *Annals of the New York Academy of Sciences* 1123 (2008): xi–xix.

O'Malley, M. A., and J. Dupre. "Fundamental Issues in Systems Biology." *Bioessays* 27 12 (2005): 1270–76.

Olby, R. C. *The Path to the Double Helix* (rev. ed.). Mineola, NY: Dover, 1994.

———. "Schrödinger's Problem: What Is Life?" *Journal of the History of Biology* 4 (1971): 119–48.

Oyama, S. *Evolution's Eye: A Systems View of the Biology-Culture Divide* (2nd, rev. and expanded ed.). Durham, NC: Duke University Press, 2000.

———. *The Ontogeny of Information: Developmental Systems and Evolution* (2nd, rev. and expanded ed.). Durham, NC: Duke University Press, 2000.

Oyama, S., P. E. Griffiths, and R. D. Gray, eds. *Cycles of Contingency: Developmental Systems and Evolution.* Cambridge, MA: MIT Press, 2001.

Pauling, L. "Schrödinger's Contribution to Chemistry and Biology." *Schrödinger: Centenary Celebration of a Polymath.* Ed. Kilmister, C. W. Cambridge, UK: Cambridge University Press, 1987. 225–33.

Perutz, M. "Erwin Schrödinger's What Is Life and Molecular Biology." *Schrödinger: Centenary Celebration of a Polymath.* Ed. Kilmister, C. W. Cambridge, UK: Cambridge University Press, 1987. 234–51.

Pisano, G. P. "Can Science Be a Business? Lessons from Biotech." *Harvard Business Review* (2006): 114–25.

Prigogine, I. *The End of Certainty. Time, Chaos, and the New Laws of Nature.* New York: Free Press, 1997.

Reiff, D. "Illness as More Than Metaphor." *New York Times*, Sunday Magazine, Dec. 4, 2005.

Rosen, R. *Essays on Life Itself.* Complexity in Ecological Systems Series. Eds. Allen, T. F. H., and D. W. Roberts. New York: Columbia University Press, 2000.

———. *Life Itself: A Comprehensive Inquiry into the Nature, Origin and Fabrication of Life.* New York: Columbia University Press, 1991.

———. "The Schrödinger Question, What Is Life?" *Essays on Life Itself.* New York: Columbia University Press, 2000. 5–32.

Sattler, R. *Bio-Philosophy: Analytic and Holistic Perspectives*. Berlin: Springer-Verlag, 1986.

Schneider, E. D., and J. J. Kay. "Order from Disorder: The Thermodynamics of Complexity in Biology." *What Is Life? The Next Fifty Years: Speculation on the Future of Biology*. Eds. Murphy, M. P., and L. A. J. O'Neill. Cambridge, UK: Cambridge University Press, 1995. 161–75.

Schneider, E. D., and D. Sagan. *Into the Cool: Energy Flow, Thermodynamics and Life*. Chicago: University of Chicago Press, 2005.

Schrödinger, E. *What Is Life?: With "Mind and Matter" and "Autobiographical Sketches."* Cambridge, UK: Cambridge University Press, 1992.

Schwartz, James. *In Pursuit of the Gene: From Darwin to DNA*. Cambridge, MA: Harvard University Press, 2008.

Scriver, C. R., and P. J. Waters. "Monogenic Traits Are Not Simple: Lessons from Phenylketonuria." *Trends in Genetics* 15 7 (1999): 267–72.

Shapiro, J. A. "Bacteria Are Small but Not Stupid: Cognition, Natural Genetic Engineering and Socio-Bacteriology." *Studies in History and Philosophy of Biological and Biomedical Sciences* 38 4 (2007): 807–19.

————. "Revisiting the Central Dogma in the 21st Century." *Annals of the New York Academy of Sciences* 1178 (2009): 6–28.

Sjoblom, T., et al. "The Consensus Coding Sequences of Human Breast and Colorectal Cancers." *Science* 314 5797 (2006): 268–74.

Strohman, R. C. "The Coming Kuhnian Revolution in Biology." *Nature Biotechnology* 15 3 (1997): 194–200.

Symonds, N., and M. Delbrück. "Schrödinger and Delbrück: Their Status in Biology." *Trends in Biochemical Sciences* 13 6 (1988): 232–34.

Thompson, E. *Mind in Life: Biology, Phenomenology, and the Sciences of Mind*. Cambridge, MA: Belknap Press of Harvard University Press, 2007.

van der Weele, C. "Images of the Genome. From Public Debates to Biology, and Back, and Forth." *Current Themes in Theoretical Biology: A Dutch Perspective*. Eds. Reydon, T. A. C., and L. Hemerik. Dordrecht, Netherlands: Springer, 2004.

van Dijck, J. *Imagenation. Popular Images of Genetics*. London: Macmillan, 1998.

von Baeyer, H. C. *Information. The New Language of Science*. Cambridge, MA: Harvard University Press, 2004.

Waters, K. C. "Molecular Genetics." *Stanford Encyclopedia of Philosophy.*
Stanford, CA: Center for the Study of Language and Information,
Stanford University, 2007. <http://plato.stanford.edu/entries/
molecular-genetics/>

Watson, J. D., et al. *Molecular Biology of the Gene* (5th ed.). San Francisco:
Benjamin Cummings and Cold Spring Harbor Laboratory, 2004.

Weatherall, D. J. "Phenotype-Genotype Relationships in Monogenic
Disease: Lessons from the Thalassaemias." *Nature Reviews Genetics*
2 4 (2001): 245–55.

Weber, A., and F. J. Varela. "Life after Kant: Natural Purposes and the
Autopoietic Foundations of Biological Individuality." *Phenomenology and the Cognitive Sciences* 1 (2002): 97–125.

Weber, B. H. "Life." *Stanford Encyclopedia of Philosophy.* Stanford, CA:
Center for the Study of Language and Information, Stanford University, 2008.<http://plato.stanford.edu/entries/molecular-genet-
ics/>

Weinberg, R. A. *Biology of Cancer.* London: Garland Science, 2006.

———. "Fewer and Fewer Oncogenes." *Cell* 30 1 (1982): 3–4.

Westerhoff, H. V., and B. O. Palsson. "The Evolution of Molecular Biology into Systems Biology." *Nature Biotechnology* 22 10 (2004):
1249–52.

Wickens, J. S. *Evolution, Thermodynamics and Evolution: Extending the
Darwinian Program.* Oxford, UK: Oxford University Press, 1987.

Wilkins, A. S. "For the Biotechnology Industry, the Penny Drops (at
Last): Genes Are Not Autonomous Agents but Function within
Networks!" *Bioessays* 29 12 (2007): 1179–81.

Woese, C. R. "A New Biology for a New Century." *Microbiology and
Molecular Biology Reviews* 68 2 (2004): 173–86.

Wood, L. D., et al. "The Genomic Landscapes of Human Breast and
Colorectal Cancers." *Science* 318 5853 (2007): 1108–13.

Yoxen, E. J. "Where Does Schrödinger's 'What Is Life?' Belong in the
History of Molecular Biology?" *History of Science* 17 35 Pt 1 (1979):
17–52.

Zimmer, C. *Microcosm. E. Coli and the New Science of Life.* New York:
Pantheon Books, 2008.

## Notes to Gumbrecht: Keeping the Singular, Risking Openness

1. From Walter Moore, *A Life of Erwin Schrödinger* (Cambridge University Press, 1994): "He despised pomp and circumstance, yet he took a childlike pleasure in honors and medals" (3f). And: "Those who knew Schrödinger divide into two distinct classes: those who considered him a person of amazing modesty, and those who thought he was one of the most conceited men they had ever met, with the majority being of the latter opinion" (48).

2. After Auguste Dick, *Einführung*, in: Erwin Schrödinger, *Mein Leben, meine Weltansicht* (P. Zsolnay, 1985), 5–10.

3. *Mein Leben, meine Weltansicht*, 11–39 (and for all subsequent page references in the text).

4. Erwin Schrödinger, *Gedichte* (Küpper, 1949).

5. Erwin Schrödinger, *"What Is Life?" with "Mind and Matter" and "Autobiographical Sketches"* (Cambridge University Press, 1992), 165–84.

6. Moore: "Erwin Schrödinger had the most complex personality of all creators of modern physics" (3).

7. Cf. my essay: "Riskantes Denken: Intellektuelle als Katalysatoren von Komplexität," in *Der kritische Blick. Über intellektuelle Tätigkeiten und Tugenden*, ed. Uwe Justus Wenzel (Fischer Taschenbuch Verlag, 2002), 140–47.

8. In the book these units are recognizable by double-spacing between successive paragraphs.

9. Cf. Moore, 143, 22–25, 17.

10. Moore: "Anny Schrödinger was an enthusiastic motorist but an inexperienced driver. About the middle of May, she saw a bright new cabriolet in a BMW show window in Berlin and fell in love with it. The Schrödingers bought the car and made plans to travel to South Tyrol" (194).

11. Cf. José M. Sanchez-Ron, "A Man of Many Worlds: Schrödinger and Spain," in *Erwin Schrödinger: Philosophy and the Birth of Quantum Mechanics*, ed. Michel Bitbol and Olivier Darrigol (Frontières, 1992), 9–22.

12. I borrow this picture and my understanding of the concept 'world' from the *kairos* chapter in Florian Klinger's Stanford University dissertation on 'Judgment'.

13. In the first part of his text *Meine Weltansicht* (cf. *Mein Leben, Meine Weltansicht*, 45–118), Schrödinger argues against the idea that the principle of physics would be "the stepwise coverage of initially insecure opinions." Instead, he calls this attitude "metaphysics." Physics, on the contrary, he associated with the "clarification and change of the philosophical vantage point" (49).

14. Cf. Moore, 188–211.

15. The published version of the Trinity Lectures was dedicated to Schrödinger's parents.

16. Moore, 31–38.

17. Moore: "Filled with a great admiration of the candid and incorruptible struggle for truth of both of them, we did not consider them irreconcilable" (35).

18. Cf. Moore, 85–88.

19. Cf. Moore, 288.

20. Cf. Moore, 141.

21. Moore, 345.